The Beaches Are Moving

Living with the Shore

SERIES EDITORS

Orrin H. Pilkey, Jr.

William J. Neal

EDITORIAL CONSULTANT

Martin Wilcox

The Beaches Are Moving

The Drowning of America's Shoreline

WITH A NEW EPILOGUE

Wallace Kaufman

and Orrin H. Pilkey, Jr.

Duke University Press

Durham, North Carolina 1983

Printed in the United States of America
on acid-free paper ∞

Library of Congress Cataloging-in-Publication Data
Kaufman, Wallace.
The beaches are moving.
(Living with the shore)
Bibliography: p.
Includes index.
1. Beaches—United States. 2. Coast changes—United States
I. Pilkey, Orrin H., 1934– . II. Title.
III. Series
GB460.A2K38 1983 333.91'7'0974 83-1761
ISBN 0-8223-0574-7 (pbk.)

ATLANTIC FLYWAY

Millions of birds:
the great tough hawks
perch on bare limbs like sleep
swans fill the bay
like a field of sea ice,
geese everywhere
so sure in their rising and falling
and gathering again.

It reminds me so much of loneliness
I am glad that four of us, friends,
watch from a hummock in a marsh
surrounded by mad certainty
so wholly impossible
so much like love.

(W.K.)

This book is for many people who have watched with us, but especially for Sharlene Pilkey, Elizabeth Anderson, and Michael Paull.

Acknowledgments

While we worked on this book we were not only researchers collecting facts and figures, but often visitors in strange places seeking friends. We cannot begin to name the hundreds of people who helped—who searched their files for us, interpreted research, and simply opened the great riches of their memories. The value of this book, or any other, does not issue from the virtues of its authors. As a writer and a scientist we are the conduits through which the people we have known may talk to others who will read this book.

Among the scientists and engineers who directed us to specific beach problems and research, we owe particular gratitude to the following: Pete Fleischer, Chris Kraft, Miles Hayes, Jeff Brown, Ralph Keer, Bob Morton, Lim Vallianos, Paul Komar, Victor Goldsmith, Stan Riggs, Bill Cleary, Paul Hosier, Francis Shepard, Gerry Kuhn, Cecil Soileau, Adrian Combe, Adrian Lawler, Jim Henry, George Oertel, Kenneth Fink, Benno Brennickmeyer, Bob Oldale, Jack Pierce, Doug Glaeser, Larry Doyle, Barry Timson, Paul Godfrey, Duncan Heron, Joe Creager, Steve Leatherman, Dean McManus, Mike Field, Dick Sternberg, Walter Pilkey, and Orrin Pilkey, Sr.

Our legal education began with the work of attorney Sarah Ramsey and continued with research done by Dan

Besse. We have always been fortunate to be able to turn to attorneys John Curry and Tom Schoenbaum with questions.

As the manuscript took shape we were fortunate to have several readers of the draft versions. Thanks to Alice Andrews, Susan Strozier, and Sharlene Pilkey. And thanks to Linda Gerber for typing what they read.

Sasha Loring's library work was resourceful and kept us moving in the final months.

Contents

Prologue 1

1 The Beaches Are Moving 12

2 Who's Afraid of Sea Level? 17

3 The Beaches Are Made of Continents 27

4 Energy: Winds, Waves, and Tides 65

5 The Beach Is Four Miles Wide 84

6 Barrier Islands: Let the Lighthouse Fall 94

7 The Human Nature of Natural Disaster 114

8 The Will to Power: An Old World Heritage 151

9 Sand Castles and Supermen 164

10 Engineered Shorelines: The Point of No Return 188

11 Who Owns the Beaches? 223

12 How to Live with a Beach 252

Epilogue: It's Not All Over Now 285

Checklist for Buying or Building on the Beach 295

Where to See the Beaches Moving 303

Bibliography of Useful References 321

Index 327

Prologue

BEACHES BECAME AN inseparable part of my life when I was seven years old and lived near a polluted, rocky scrap of coast constantly littered with debris and mined with old bottles and rusty metal. It was and still is a sad beach in a proud and handsome town.

The Incorporated Village of Sea Cliff is an hour's train ride from New York City on the hilly north shore of Long Island. It is a fashionable town of white-collar commuters, boutiques, craft shops, art galleries, and restored gingerbread houses. Thirty years ago when I was growing up, it was a town of decaying and boarded-up hotels, empty storefronts, cheap old houses, and five thousand people as common as anywhere in the United States. My father was a factory worker who returned home from Brooklyn every night with grime black in the pores of his hands and face. Few common people can afford to move to Sea Cliff now, and many who once lived there have cashed in on the vogue for old homes and gone. Something has remained steady through all the change. It is something that reveals in all of us—past and present residents—a single mysterious and mystified humanity.

Everyone in Sea Cliff loves the beach. All 118 miles of Long Island and this beach were dropped in place some twelve thousand years ago when the last glacier began to retreat and left behind a great ridge of topsoil, rock, and sand scraped from the face of New England. Most of Sea Cliff sits high above a fingerlike harbor, and from a little park at the highest point of the bluffs you can see almost the entire shoreline. For as long as I can remember the harbor has been flanked by homes, sand quarries, incinerators, marinas, and factories. The water I learned to swim in and contemplated for five years as a lifeguard has been red, white, purple, and brown with pollution. In the first-aid room I bandaged innumerable feet torn open on glass and metal in the sand and underwater.

Today the shores are more crowded than ever and the water is almost never clear. To dig clams with bare feet as we used to do is utter folly. Below and above the waterline the beach glistens with broken glass. What we really loved about the beach, however, is still there. The tides come and go. The calmest weather yields the faint sound of water's endless motion. There is a sense of space and more seas beyond.

If you look north up the harbor beyond the half-mile-long rock jetty built by J. P. Morgan, you see Long Island Sound. Beyond it, though unseen, I could always imagine the Atlantic Ocean, the Arctic, and the Antarctic oceans. With each additional reach of water I could imagine a more powerful sea. The tides and waves of the harbor are only the smallest expression of forces whose power and size are more intriguing to imagine from the top of an old bluff than to see firsthand from the bare ocean beaches or from the deck of a ship surrounded by nothing but water.

There is something mysterious and awesome about seeing the sea or any portion of it from the midst of human structures, machines, noises, and motions. It is like having the most untamed wilderness in your backyard, or the sensitive tip of a tentacle of an unpredictable wild beast resting at your feet.

Why else is it that when we are most bound up and tortured by the squabbles and disasters of life, confronting the water seems to lend perspective? Why else do I remember that almost every couple in love, both the infatuated young and old feeling renewed love, would go walking along the beach? Isn't love our most intense, most mysterious and unpredictable emotion? Isn't it the one which seems to set us apart from our neighbors and defy the claims of civilization?

Almost everyone in town detoured once or twice a month on the way in or out of town to drive along the harbor road. Any mild day in winter, spring, or fall at lunchtime you would see a dozen or so cars and trucks parked opposite the water—deliverymen, plumbers, linesmen, factory workers, hardhats, and businessmen eating a sandwich and gazing at the water.

I don't know exactly any one person's reasons, and I don't even understand my own, yet until I left to go to college I went to the beach more often than anyone else in town. And I sought out the most private stretches.

The harbor road parallels the beach for only a few hundred yards, and for almost a mile the steep glacial sand bluffs come right down to the high-tide line, or at least to the seawalls and bulkheads that are supposed to keep the bluffs from sliding and the houses on top of them from falling.

Along most of this mile few people swim. To walk it you have to climb over a hundred little rock and timber walls jutting out into the water from the bluff. These are supposed to keep the beach in place. But there is no beach at all at high tide, and at low tide the shore is strewn with the rocks that have washed out of the bluffs. Most people use the town's beach, where smoother sand is periodically dumped at great expense.

After I learned to swim I spent the summer days diving for clams and digging steamers and bait worms. There was some money in this. At high tide I strapped on fins, mask, and snorkel and went spearing black fish and eels in the rock walls, using an old pool cue tipped with a steel head and propelled by a band of inner tube tied to the handle. During the winter I cracked the red hematite stones to find fossils of willow leaves from a time before the glacier had made Long Island and when New England was semitropical. On and off for four years I worked with a local archeologist excavating a shell mound on the bluff a half mile from Morgan's jetty. This was a rare site once inhabited by Indians who lived before the invention of pottery in North America. Unfortunately, half of the site had collapsed into the water with the bluff, and more disappeared even while we dug. When it was too cold to do anything else, I would just wander along the beach, my face freezing in the spray and my hair getting stiff with salt. In the wind, the water, the rocks, and even the ruins of old boathouses and boardwalks, I tried to see my own moods. Over the years I contemplated every inch of the beach. I never could fool myself into thinking nature sympathized for a moment. That indifference was yet one more torture for adolescence. In the pain I learned things about the beach which I did not know I knew until a few years ago.

Had I been born ten or fifteen years earlier, I could never have enjoyed so much privacy. I could never have had such loneliness in which to quench my own loneliness. I could have had nothing so wild to love with my own wild desire to be loved. From the late 1800s to 1938 there had been a wooden boardwalk along the entire length of the beach, and on the boardwalk scores of shops and Tilley's steamboat dock with daily departures for Connecticut. The Village of Sea Cliff has been dotted with huge gingerbread hotels, spacious boardinghouses, and fine restaurants.

In 1938 a hurricane crossed Long Island. From New Jersey to New England a great wind and tide destroyed everything along the beaches. In our town the harbor swelled with green waves as tall as houses, and when they subsided nothing was left of the boardwalk except jagged pilings and driftwood.

I heard the story often when I was young, and secretly rejoiced that nature had freed the beach and the bluffs from all those adults. I liked having the beach for myself and my friends. Small bands of kids raged up and down the beach in mock wars. The wooded, brushy bluffs we called the "Eighteen Trails," though some claimed there were only fifteen and others said twenty. In the soft sandy hillside we dug small rooms and lined them with driftwood. Through gunslits we could survey the harbor. We tied ropes to the limbs of the big oaks and had swings that arched out over the high-tide line. If you were crazy, you could drop to the water. If you were sane, you dropped to the soft hillside.

Kids wore out the section of bluffs we used most. When I returned in 1963, after college and graduate school, I saw that the big oaks had fallen, the sand and gravel gradually pushed from beneath them by hundreds of sliding, jumping kids. The trails had eroded and large sections of the brush hillside had slipped downward, leaving behind a blank white face.

I was ashamed at the destruction in which I had happily participated. Yet I learned some years later that I was not the only one who had helped destroy the natural processes of the shoreline.

The archeological site which was washing into the sea was a victim of currents altered by Morgan's jetty. The rocky beaches were caused by an interaction of eroding waves and bulkheads holding back new sand from the bluffs. The good spearfishing in the rock walls was a result of the walls causing wave erosion on one side and thus providing deepwater and exposing nooks and crannies for eels and fish to hide in as they awaited smaller fish which fed among the rocks. The homeowners on top of the bluffs who were spending millions of dollars shoring up their houses were the victims of kids destroying the vegetation, and they in turn tried to terrace the bluffs to lock up the remaining sand. Finally the economic collapse of the town in the 1940s, which enabled a poor family like ours to buy a solid two-story house for four thousand dollars, was caused by the sum total of mankind's arrogance in investing so much private and public money in development that literally hovered over the high-tide line.

What lay behind all these problems was the wild, dynamic motion of the sea, which was the very reason why everyone in town loved the beach. Actually, we loved the water and ignored the beach. We made the mistake of assuming that the sea and the land bordering it were two entirely different domains. We saw no connection at all between what we did on the beach and what the sea seemed to do to the beach. We felt betrayed when the water undermined a piece of the bluff or when a storm washed out piers, seawalls, boathouses, and archeological sites.

When I moved to North Carolina, where there are more miles of undeveloped coast than anywhere in the East, I became a Director of the Conservation Council of North Carolina. To provide the Council with expert testimony at public hearings I approached a number of scientists at Duke, the University of North Carolina, and North Carolina State University. This was in 1968, two years before the first and biggest Earth Day celebration made environmental activism respectable. Many scientists felt it was unprofessional to appear publicly on behalf of amateurs fighting for a cause. Others more bluntly said it would cost them consulting fees, and their graduate students would be cut off from jobs. The clearest exception was a marine geologist at Duke who, despite warnings and criticism from colleagues, volunteered to testify in favor of a mild bill to regulate open-pit mining. After his testimony one of his senior colleagues issued several apologies to other geologists.

Orrin Pilkey is not easily scared. He became a geologist only after serving with a Montana parachute team of firefighters. He is still an aggressive touch-football player and formidable on the basketball court even at five foot six. He runs the Marathon in Boston and wherever else he has time. He was not to be stopped in the first leg of his public career by academic politics.

Pilkey was already well-known for his study of seafloor sediments when he first became interested in beaches. A dead calm at sea and a giant hurricane combined to turn his attention from the dark seafloor to the bright beach sands. In August 1969, Hurricane Camille smashed into the Gulf coast of Alabama and Mississippi. A day later, Pilkey flew to Waveland, Mississippi, to help his retired parents shovel the mud out of their new home and see if they could restore the house. They were more fortunate than the owners of the two rows of houses between them and the Gulf. The waterfront row was beaten into matchsticks and rubble. The second row was knocked from its foundations and partially destroyed. As Pilkey cut trees and shoveled mud out of the living room he remembered how two months earlier his father, a structural engineer, had driven him around the area pointing out examples of how not to build and how to build in a hurricane-prone area. Later when he looked for the "stormproof" houses, he found almost nothing but a few poles and sticks. On the beach a major storm spares not even the best, he concluded. The Navy Seabees were sifting through the wreckage of both the right and the wrong homes, clubs held ready against snakes as they searched for bodies.

A few months later, Pilkey was aboard the research vessel *Eastward* with Smithsonian geologist Jack Pierce. The Atlantic had been dead calm for ten days, and the technicians taking tedious bottom photographs were finding nothing of great interest. As Pierce and Pilkey played poker in their long hours off, Pierce started talking about his recent studies of the geology of North Carolina's Outer Banks. He was turning up more and more evidence that the islands were changing not only their shape but also their very position on the map. In passing, he noted how often he saw residents trying to prop up their houses against the advancing ocean, or ward off the waves of a migrating inlet by building rock groins out into the surf, or throwing up makeshift dunes—small efforts leading to larger efforts, just as losing poker players will stake their all rather than quit.

The fresh memory of Hurricane Camille and his parents' narrow escape made Pilkey brood on man's misunderstanding of beaches. He began thinking about other beaches he had known, especially ones that had been settled for many years. The different seascapes shared a common element—ample evidence of man's defenses washed away. The older beaches seemed to be disappearing themselves, replaced by more and more massive defense works—seawalls, bulkheads, jetties, and revetments.

While working at the Duke Marine Lab on the North Carolina coast and consulting with the Conservation

Council on coastal development, Pilkey became increasingly worried that the long wild stretches of barrier islands and the state's comfortable little coastal villages would soon begin to look like the coast of New Jersey. It was not so much the population, motels, and subdivisions that worried him as the growing potential for disasters even greater than the famous Galveston hurricane of 1900, which took six thousand lives in one night.

Pilkey started to read more and more of the new literature on beach formation and history. He saw clearly that the public and the government that protected the public were laboring under a fundamental misunderstanding of the beach. To be healthy, beaches and sometimes whole islands have to move. Trying to "stabilize" the beach was like trying to stabilize the ocean itself. Old King Canute had retreated when the waves would not stand still at his command, but Americans were responding by throwing more and more technology at the sea. The most alarming piece of literature Pilkey read was the *National Shoreline Study* published in 1970 by the U. S. Army Corps of Engineers. The report proposed almost two billion dollars of engineering works to protect "endangered" shorelines.

Pilkey noted that most of the endangered shoreline was located amid heavy development and near former stabilization work—much of it undertaken, not much earlier, by the Corps itself. With a scientist's passion for verifying generalities and theory by observation of particulars, Pilkey began studying Bogue Bank, a barrier island a few miles from the Marine Lab. It was here that he gave me a crash course on beach problems—geological, historical, and political. Bogue is a developing barrier island, much of it once the property of Teddy Roosevelt's family. There facing the Atlantic were all the mistakes, and more, that I had seen in my own quiet harbor. Dunes had been bulldozed, inlet paths built upon, jetties and groins pushed out into the water to trap sand, bulkheads squared off against the ocean, and trees cut to enhance the view and provide parking. At the east end old development problems were represented by Robert E. Lee's Fort Macon, its ramparts refortified against the sea by jetties. At mid-island the Roosevelt estate was rapidly passing from the hands of the conservationist's heirs to the hands of the subdividers. At the western tip we watched the developer's paved road falling into a migrating inlet. Most of the land claimed by the inlet had just been donated by a prominent banker to a private school for a tax deduction. The remaining lots were being heavily advertised and exhibited by local realtors.

As a realtor and appraiser myself, I was unhappy with what I saw—yet more food for the public's distrust of the developer and real estate broker. I went to the beautiful old brick courthouse in Beaufort to see who was doing all this and for how much. The deeds read like a *Who's Who* among the state's well-to-do and politically influential citizens, not to mention local politicians. The documents were affixed with revenue stamps proving the millions being made at the island's expense. Some leases, trades, and transfers indicated the usual tax evasion through worthless gifts and bogus rentals, but most of the business was technically legal and ordinary enough.

While I was busy lobbying for a Coastal Area Management Act, Pilkey was writing a little guidebook to Bogue Bank: *How to Live with an Island.* It was an auto tour of sound and unsound development keyed to the car's odometer by tenths of a mile. It named those high-rise motels which would topple in the next major storm, and located the sections where waves would overwash subdivisions. It became a local best-seller.

Both Pilkey and I wanted to do more, to provide a picture as large as the Corps' own *National Shoreline Study.* We believe that study has set the stage for the most massive technological defense of coastal development ever attempted in the world. Much of the development it will encourage and protect will become part of storm disasters larger than any this nation has ever known. In other areas, where development survives behind its fortifications, the beaches will disappear. The struggle to "defend" the coast is the acting out of an understandable human sentimentality, hedonism, and faith in technology. This book is as much an exploration of those human traits as of the beaches themselves. It is our explanation of how we have destroyed so much of a resource that has tremendous value to both the economy and spirit. Most important, this book is our attempt to offer a vision of how we can live with and love our beaches without destroying them or ourselves.

Wallace Kaufman

1

The Beaches Are Moving

A PERSON CAN live on a mountain, in a desert, or on a suburban lot, and after a few years he can say: "I know every inch of this place, every rock, tree, weed, and fold of earth." No one can say that about the shoreline or the beaches.

The beach is land which has given itself up to wind and wave. Every day throughout the life of the earth, the wind and the waves have been at work shaping and reshaping the beach, pushing and pulling almost microscopic grains of sand and sometimes boulders larger than cars. For practical purposes there may be days when no wind blows, but even on the stillest days, when the sea lies like a flawless mirror, a wave moves against the shore. This is the wave of the tide, a gigantic slow sloshing of the sea within the oceanic basins. As the sea and the wind move, the beaches move.

A glance at maps of the American coastline, from Vespucci to the U. S. Geological Survey, seems to show little change. There's the Florida peninsula, the familiar elbow of Cape Hatteras, the great fish form of Long Island, and the hook of Cape Cod. Yet, comparisons of detailed maps show that privateers, pirates, and adventurers sailed a different coastline than our Founding Fathers defended against Britain. The coast illumined by the "rockets' red glare" is not the coast of our Bicentennial. From map to map islands change shape, inlets appear and disappear, shoals shrink or grow, and capes of land stretch further and further out to sea. To know the beaches is to know the beaches are moving.

We ignore this when we build motels, pavilions, boardwalks, and even whole towns on the edge of the ocean. In our business hats we do not recognize any real estate as movable. Corners are staked, lines drawn, and neat rectangular lots are recorded in courthouses as if they would be true forever. But in our play we are guided by a different vision. On almost any summer day on any sandy beach in the United States, adults and children alike amuse themselves with sand castles, moats, road systems, arches, miniature towns, and statues of each other. Here at the beach is one of the few socially acceptable forms of carefree play for adults. In molded damp sand we build images of our dreams and hopes. The water takes them away, as time one day takes our very lives. Amateur engineers may be fascinated with what we build, but perhaps the more interesting question is, Why do we make these little sacrifices? Perhaps, freed of economic and social concerns, we set aside our passion for monuments, history, and permanence and admit we are part of the ceaseless change that is nature.

The waves and tides which accept our sacrifices of sand castles and statues are only a small part of the forces that keep the beaches in motion. One way of understanding the dynamic nature of the beach would be to imagine yourself a photographer setting out to do a photo essay and a film on the beach. An ordinary snapshot would reveal a static, stationary place. The picture would show no sign of motion unless it were taken on a day when wind, rain, or waves were so violently agitated that they registered as blurs on the film. Now suppose you go to the water's edge and snap a picture at an extremely fast speed, say 1/2000th of a second, and on very fine film. And you do the same at the top of a dune. Very likely the pictures would show thousands of minute sand grains tumbling in the water and blowing through the air. Finally you set up a small shelter and a time-lapse camera and take two or three pictures a day for several years. Then the pictures are shown like a movie. The film will show dunes rising, falling, and marching along the beach as if enormous whales were migrating beneath the sand. The foredune and the tidal area of the beach where bathers spread their blankets will expand and contract like a sheet of stretched rubber. At the same time the tidal zone as well as the beach seems to rise and fall like the gangplank of a boat riding the daily tides beside the dock. The sand of the beach may even turn miraculously into large cobbles, then back into sand, and so on. During stormy weather changes flash before us almost too fast for the eye to follow—a dune or an inlet here one second and gone the next, as the film moves on. Chances are that at some point in the film we would see a dark line cut across the beach not far from the camera—an inlet. It may advance in our direction, gnawing away at the beach before it and leaving new beach in its wake. Within a few minutes the film will stop short, the photographer having abandoned the camera as the whole lookout station collapses in the water.

Despite this incessant motion, beaches continue to border the continent with about the same area from one year to the next. But like a person constantly changing position in a large armchair, not everything will be in the same place all the time.

Beaches are not stable, but they are in *dynamic equilibrium.*

We say the site of New York City or Chicago or Atlanta is stable because even over several human lifespans there is little practical change in the land mass itself unless mankind so wills it. Dynamic equilibrium is not stability of this kind but a net balance among many changes. If you give a person a dollar in return for three loaves of bread, you have neither a net loss nor gain. Your assets change shape but their value is in equilibrium. The same is true if you eat the bread. You no longer have the loaves, but you have energy. You have been making "trade-offs." You trade off the convenient convertibility of stock for the comfort of a house, or you trade off the supply of bread for more physical stamina.

Nature maintains the dynamic equilibrium of the beaches through unceasing trade-offs of four factors:

MATERIALS: sand, silt, biological debris, flotsam.

ENERGY: the forces of winds, waves, and tides.

SHAPE OF THE BEACH: steepness and width.

SEA LEVEL: land rising or falling, seas gaining or losing water.

The simplicity of listing these four elements is deceptive. Great mysteries still surround the processes of shoreline evolution. We may never unravel the interworkings of the long slow rise of sea level and the daily work of wind, wave, and tide. Yet gradually we are learning how the beaches survive, so that the story of a particular event, limited to a short stretch of coast and span of time, can be told.

Today we understand how hurricane seas surrounded and leveled the island city of Galveston and killed six thousand people in 1900. We know how a resort town disappeared on Oregon's Tillamook Spit. We understand the evolution of Cape May, New Jersey, from a broad sandy beach resort frequented by the country's Founding Fathers in the eighteenth century to a rubble-walled town pleading for economic assistance from Washington. We can prove that Miami's once broad white beaches were destroyed by the hotels facing them. We know how in an Alaskan bay giant waves sheared trees from hills a thousand feet above the water. We can even predict where future disasters will occur.

Beach stories are too easy to find. In writing this book we amassed whole file cabinets full of stories. Some stories were a thousand years old—stories of native American coastal villages and beach encampments only revealed in this century by a changing shoreline or archeologist's trowel. Most of our stories are modern because today the number of people living within a mile of the beach is probably greater than the number of all Americans who lived anywhere in the nineteenth century. We can predict the headline stories for the next decade or two because mankind's presence and the elements of the dynamic equilibrium that write these stories will not have changed. We are tempted to tell too many of these stories because they embody so much spectacular action, so much complex human character, and, at least at a distance, so much amusing irony.

Yet accounts of houses falling over California cliffs, or Gulf Coast islands disappearing in hurricanes, are in themselves unimportant. Only idle curiosity wants to know how high the flood was and how many perished. Our higher intelligence asks: *How did we face danger? How did we feel afterward? Why did we take the risk? What are we going to do now?*

2

Who's Afraid of Sea Level?

FROM TIME TO time an article appears in the newspaper describing the various disasters which might end the world—collision with a comet, the moon tiring and falling out of orbit, the sun exploding or cooling, or the earth being drawn into the sun. Apocalyptical disaster can be fascinating. With the exception of a few people, however, motivated by an unusual prophetic perspective, no one's life is changed much, if at all, by the sure knowledge that the world will come to a spectacular end.

To those who contemplate it, rising sea level enjoys a status similar to the other visions of Apocalypse. We know the oceans rise and fall. We know the slow wrinkling of continents raises and lowers the earth itself. We acknowledge the marine fossils of the walls of the Grand Canyon and New York's Catskill Mountains. But it all happened too long ago to worry us by its return. There are no human bones in the Grand Canyon's fossil beds. Most people are unaware of whether sea level is rising, falling, or standing still. Those of us who have lived a few decades might dismiss any thought of danger by observing that in our lives we haven't heard about the ocean's rising into the streets. What have we heard? Something about a few hundredths of an inch a year? Certainly neither New York nor San Francisco will follow Atlantis, at least not for several centuries or millennia—unless, of course, air pollution really does create the "greenhouse" effect and the icecaps melt completely. Even without accelerated melting of icecaps some shores will drown before others. Rising sea level could cause the destruction of Galveston County, Texas before the year 2000. The water is already in the streets. In Bay City the fish in Henry's Fish Market are almost back in the sea, and the buyers at high tide are walking in the water as they shop. You will understand why when we have explained the character of sea level.

If we think of rising sea level as merely the addition of water to the ocean basins from the continents and melting icecaps, we do not have the whole story, but we do have one fundamental cause. During the greatest expanse of the last glacier so much water was locked up as ice that sea level fell to four hundred feet below our present shoreline. Divers, research vessels, and remote-control cameras have filled laboratories with evidence that under the waves a hundred miles seaward from New York or Boston fertile plains and valleys once grew lush and dry. There is no other way to explain beach rocks, coral reefs, salt-marsh peat, and oysters on the deep ocean floor. Much of the landscape trekked and hunted over by our cave-dwelling ancestors nineteen thousand years ago is now flooded.

Since then sea level has risen, but geologists argue about how steady the rise has been—whether there were pauses and during what periods the rise was most rapid. In general, the rise seems to have slowed down about five thousand years ago. Since that time the rise in sea level may have speeded up, paused, or even fallen for brief periods. For the first part of this century sea level was rising very slowly. Then in 1971, from the National Oceanographic Survey, Dr. Stacey Hicks, a physical oceanographer, published evidence that the rise of sea level had begun to speed up. Measuring against the New England coastline, he documented an increase from one foot a century to three feet a century. Even that dramatic increase amounts to only three inches every ten years. What's more, this figure is a generality full of exceptions. In some places sea level is rising much faster, and in other places it is falling. Anyone who knows much about water knows it always seeks its own level, and that of its own accord the sea can't be higher in Massachusetts than in Alabama. In fact it is.

One problem of cutting a sea-level canal across the Central American isthmus would be a current running continually from Pacific to Atlantic, sweeping along alien plants and animals. Sea level is higher on the Pacific side of the Panama Canal. Here, as all over the world, the sea's level is affected by winds, tides, weather trends, currents, and even water density and temperature. The effects of wind may last only a few days, while a current as steady as the Gulf Stream or Japan Current may push water shoreward for many years. So sea level has many causes. Sea level is not an absolute measure, a uniform horizontal surface, like water level in a bucket. We can only measure sea level as it exists in a given locale. And we can only measure it relative to other things which may themselves be in motion. The land against which we measure sea level moves. Canada and upper New England are rebounding from the weight of the glaciers. All of the great crustal plates of the earth are in motion, rising and falling as they move against each other. Parts of the floor of the Gulf of Mexico are sinking under the weight of the Mississippi's delta. Little, if any, of these motions are evident to the observer who notices only that the water is higher or lower on the cliffs, beaches, bridges, piers, and seawalls.

Steady, long-term changes in sea level can usually be explained by one of three events: more water entering the oceans; the seafloor rising or sinking; and the edge of the continent rising or sinking. If only we could measure the volume of water in the oceans, we would have a steady reference point and we might know which of the three events is changing sea level. With a steady volume local differences in sea level would clearly indicate motions of the continental plates. A worldwide rise in sea level would mean either the unlikely syncopated falling in the motions of all the continents, or a rising of the seafloor. The latter would act something like a dent or growth in the bottom of a bathtub.

Most geologists agree, in spite of the measuring difficulties, that measured against most landmarks, sea level is generally on the rise. They disagree about the causes. A minority argue that the seafloor is rising in places, displacing water. Every ocean has a mid-ocean ridge where the earth's crust is spreading apart. The Mid-Atlantic Ridge, for instance, is so active that recently volcanic ruptures created Surtsey, a new island off the coast of Iceland. Growth in one or more mid-ocean ridges, or an immeasurably slight bulge over hundreds of thousands of square miles of seafloor, would displace enough water to create a significant rise in sea level.

Most geologists think the polar icecaps are still melting.

Thousands of years of precipitation are still stored around both poles.

The arguments over cause are interesting, but not so important to coastal residents as the simple fact that the water is rising. Should you prepare for the floods? The answer may depend on where you live.

The rates cited by Dr. Stacey Hicks are only general, and in some places sea level is rising much faster and people are responsible. Along the northeast Texas coast underground shale layers are expanded by thin layers of trapped water. To slake the thirst of cities around Galveston Bay, including Houston, Galveston, and Texas City, local governments have been pumping some six hundred million gallons of this fresh water a day. As the water leaves, the shale collapses and the land sinks. At some point the rock will be compressed to stability, but by that time the rise in sea level may turn the space city of Houston into a sea city. Over the past fifty years or so some thirteen hundred square miles of coastal plain in Texas has subsided or sunk more than one foot.

Many large, lavishly built homes near Baytown have protected themselves with expensive flood walls. Even so, storms now flood the streets and hold the residents prisoners in their own homes. Some homes are now permanently surrounded by water and have been abandoned. The Humble Oil Refinery, a major user of the ground water, protected itself in 1965 with a 13.5-foot-high levee. Already storm water tops the structure. At San Jacinto Battleground Park a sign which says NO FISHING FROM BRIDGE is near a lot of water, but the bridge is out of sight—submerged. The San Jacinto monument no longer towers higher than the Washington monument. In some places subsidence has reached nine feet.

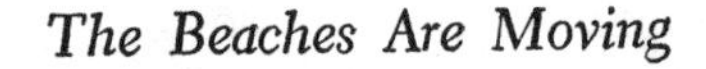

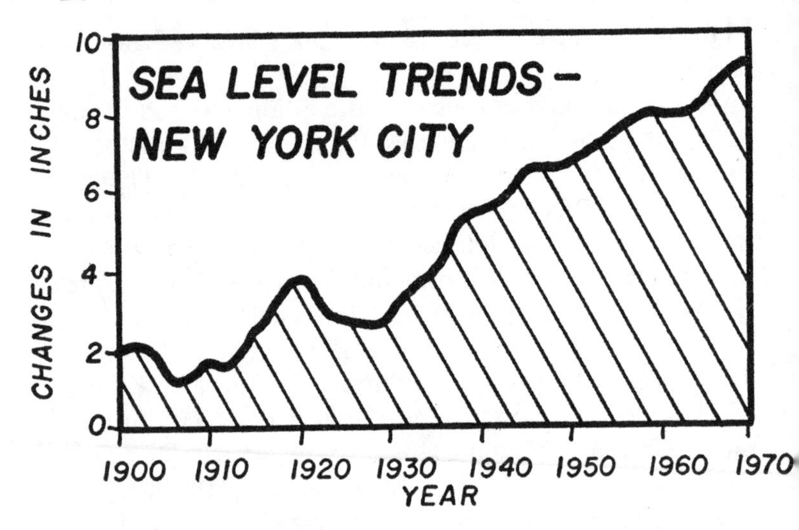

Readings taken from very accurate tidal gauges in New York City show the relentless sea level rise that is behind the inevitable erosion of the world's beaches.

To the house on the shore, feeling no fall, the land sinking means the sea is rising. Geologists estimate that by the year 2000 this sinking of the land and rising sea level will mean a loss of some twenty thousand acres of shoreline property. But even this year or next, in a hurricane equal to 1961's Carla, high water and waves, rising on top of the higher sea level, may flood seventy square miles of land that stayed dry a generation ago. Here is the real importance of rising sea level—a small rise may bring water far inland.

From Texas to New England many miles of coast are within a few feet of sea level. The upward slope of the coastal plain is often only one fiftieth of one degree, or less than two feet in a mile. The coast has protected itself by building beaches and dunes to confront the sea. Standing at the tide line and looking up the beach and toward the tops of the dunes, we might be tempted to think that, even at three feet a century, the rise in sea level would take several lifetimes to make significant changes.

A few years ago a Virginia Beach newsman decided to amuse his readers by applying *reductio ad absurdum* to the "alarmist" statements of geologists about the dangers of rising sea level. He calculated the date at which water would reach the average elevation of the city's residential and commercial areas. He concluded that everyone had better be ready to move by the year 4000. By that time, however, Virginia Beach will be several miles out to sea. The reporter would not have been so glib if he had been familiar with the history of Mr. E. R. Outlaw's house only forty miles south in Nags Head, North Carolina. The Outlaw house has had to be moved three times in the past century. Each time it has been moved some 200 feet back from the advancing shoreline. Thus, $3 \times 200' = 600'$ a century; divided by 100 years, it equals a shoreline retreat of 6 feet a year. When sea level rises a few inches the shoreline may retreat many feet, and seemingly solid cliffs may suddenly collapse.

If the beaches and dunes were formed of granite or cast of concrete, we might live out our lives behind them seeing the sea level slowly creep up the solid wall with only an occasional wave swash or storm surge overtopping our protection. (Provided the walls were also rooted deeply below the ground and invulnerable to freezing, pounding, abrasion, etc.)

But let us step back a minute and look at the beaches in the perspective of the entire coastal plain. On that incredibly flat surface the beach ridges and dune systems are very small, narrow irregularities, standing between the land and the vast ocean. In most places the protective dune system averages only five to ten feet high, lower than most storm waves. And this wall, which beach communities often rely on for defense, is made of sand and meant to move. The dunes along the beach are not, and cannot be, a dike.

The nature of sand is to move. That is how it came to the beaches, how it blows from the beaches into the dunes, and how it washes from the dunes into the bays to form new salt marsh and forest. Along the Gulf and East coasts the sandy beaches do not turn back the sea; they defend by strategic retreat.

Those who live near the shore choose to say that a shoreline moving with the water toward their house is "eroding." The U. S. Army Corps of Engineers, charged with defending the beaches, also uses the word. So do geologists. But when they say "erosion" they do not wave the word like a battle flag unless they are employed by an engineering firm. Most geologists speak of beaches *retreating*. Barrier islands are said to *migrate*. Beach erosion, geologically speaking, is not usually a permanent loss, but a strategic retreat. As any war-game player knows, fewer troops are lost in strategic retreat than in unending battle. The barrier islands which protect most of the Gulf and East coasts have existed continuously for thousands of years, but they have retreated many miles. They are warehouses of sand supplied by ancient rivers that no longer deliver directly to the ocean beaches. To preserve this sand the islands migrate up the coastal plain, picking up sand even now being deposited by rivers in the estuaries and bays. Behind the islands the shore of the mainland also retreats. How fast this happens depends on the slope of the coastal plain. The more gentle the slope, the faster the island and its shoreline retreat. Since the end of the last glacial advance along parts of the Gulf Coast, sea level has at times pushed the shoreline back one hundred feet a year. In the coastal plain states a one-inch rise in sea level means a horizontal shoreline retreat of a hundred to a thousand times that distance. Perhaps we do not notice the retreat on natural beaches because, despite the motion, the beaches from year to year seem unchanging. This is Brunn's Rule, which says that as sea level rises the retreating beach assumes exactly the same shape in a new position.

The faces of the seacoast and the actions of the oceans are so varied that until recently people blamed shoreline erosion on weather patterns, ground-water flow, storms, or tides, as it suited their perspective or politics. A few years ago, while still a student at the Virginia Institute of Marine Sciences, Dr. Peter Rosen studied the varying rates of sea-level rise around Chesapeake Bay. Through a skillful use of statistics he proved that the rate of shoreline erosion is everywhere controlled by the rise of sea level. Rosen's study stripped away the many masks that have led scientists and laymen to blame erosion on forces that seemed more susceptible to human control.

The role of sea level is clearest on the sandy beaches from Long Island to Texas, where the retreat of the low beaches on a low coastal plain is steady and measured with relative ease. On the rocky West Coast, on the mangrove coast of Florida, and on the glacial deltas of Alaska, it is harder to see. Yet sea level is rising the world around, and almost everywhere the shoreline is retreating. Those coastal residents who like company in their misery should know that (with a few local exceptions) the whole world is eroding. This is true even of the great Pacific sea cliffs and the seemingly invincible rock of New England headlands.

If sea level were falling around the world, we would have few coastal erosion problems. With sea level rising, all our efforts to control erosion are temporary. Nature has her own defenses. Unfortunately for those of us who crave the comfort of fixed property lines, those defenses all involve motion. How fast that movement is will determine the cost and viability of mankind's defenses. The speed of erosion depends in large measure on another part of the dynamic equilibrium—*material,* or what the beaches are made of.

3

The Beaches Are Made of Continents

ALTHOUGH THE SUNBATHER lying sleepy and vulnerable on the warm sand seldom knows it, he is lying on one of the most ingenious defensive structures in the world. In nature's endless interplay of force and material, the beach is a buffer zone, shock absorber, and biological way station between the sea and the land. All this is made of wastes, materials cast off by the continent, and in some cases by the life under the sea itself. In nature, of course, waste is mainly a matter of definition and point of view. The wastes that form the beaches are not only vital in the coastal environment but are the first line of defense for the mainland from whence they issued.

Since the ancient separation of the land from the water, exposed rocks have been splitting and crumbling, forming smaller and smaller pieces and finally cobbles, pebbles, sand, and mud. Some of this debris becomes a vital part of life-giving soil. Sooner or later most of it reaches the edge of the continent.

At this moment, high in the Cascade Mountains of Washington, a mountainside lies far beneath a thick sheet of ice, a glacier moving with infinite slowness. In the arid Badlands of South Dakota groundhogs excavate their burrows in weathered soil. These animals, as averse to extra work as humans, always push the excavated material downhill. High in Minnesota a cold clear ribbon of water, the beginning of the Mississippi River, slips over rock and curves against low soil banks. In Alaska's Tongass Forest a spruce loses its rocky footing and falls across the stream, loosening a few pebbles from the earth. So for many millions of years in the most remote parts of the continents, the materials of the beach have begun their journey toward the sea.

The sand, silt, debris, and flotsam which form the beaches of America come also from the Appalachian Mountains, the Grand Canyon, the glacial bluffs of New England, the bays and estuaries of New Jersey, the coral reefs of Florida, and the dark continental shelves beneath the Gulf of Mexico and the Atlantic and Pacific oceans.

The story of beach materials is even more complex than this, and many of us understand it no better than the king who lost the recipe for his favorite muffins understood baking. He commanded his wise men to analyze the last muffin and say of what it was made. They brought back their findings: the muffins were made out of crumbs.

To say that beaches are made of sand does not tell us how to make a beach. The recipes for beach material are often simple, but there is no national cuisine. Each beach has its own special materials, its own provincial recipe. Nature relentlessly reminds us of this even in relatively quiet inland waters.

In 1955 the U. S. Army Corps of Engineers came to rescue the eroding beaches of the Presque Isle peninsula in Lake Erie. From the quiet harbor behind the curving peninsula huge pumps poured 150,000 cubic yards of sand a year onto the lakeshore beaches. For a while bathers enjoyed broad, soft shores, but the sand soon began to vanish. Four years later the beaches were still losing 25,000 cubic yards a year. In the 1960s the Corps returned and pumped in 688,000 cubic yards of sand, enough to fill several football stadiums to the top of the stands. Again the sand disappeared.

The part of the recipe overlooked was grain size. The sand the Corps had been pumping out of the harbor had smaller grains than the natural beach. A beach of small, closely packed grains is not as porous as a coarser beach and does not absorb as large a part of incoming waves. The combination of stronger backwash and lighter grains quickly washed out the public investment in saving the beach. The money is gone with the waves, but the beach was finally saved when the Corps changed tactics and pumped up 1.2 million cubic yards of coarse sand.

Not only does beach material vary in size from microscopic grains to boulders, but the recipe varies with the season. Each beach gets the materials it needs to survive under a given set of conditions. An expert studying the materials of the beach on any given day can tell us something about the mood of the sea which transported the material. The beach preserves itself against seasonal changes in the sea much as humans preserve themselves by dressing according to the weather. During storms, and in fall and winter when waves roll in faster and higher, more and more fine sand is captured and held by the surf. It only comes to rest offshore in deeper water. The larger, heavier sand grains and pebbles remain on the beach. The coarser materials of the winter or storm beach, both above and below high water, act like a battlefield full of pitfalls over which an onrushing army advances and retreats, losing some of its force with each motion. The coarse sands are in fact a kind of natural armor. But fine or coarse, the secret of success is adaptation and flexibility. Beaches of sand or gravel can absorb blows that tear apart concrete walls in a few seasons.

Even driftwood can be a valuable beach material. The beaches of Grand Isle, Louisiana, are forever covered with bits and pieces of tree limbs from inland forests and flotsam from the Gulf. In the 1930s bathers convinced local officials to clean up the beach. No sooner was it clean than it began to disappear. Day by day it grew smaller, and during storms erosion would reshape the entire length of the beach. Prior to 1933 and the anti-driftwood campaign, erosion had been slow. The driftwood detested by bathers was not just debris, it was part of the beach itself, and when it went trouble followed. Though driftwood moves with the tides, it also seems to be effective in breaking waves and trapping sand—a kind of natural seawall and sieve, and no less a part of the beach than sand.

As singular as beaches are, their life depends on the sea in front of them and the landscape for miles behind them, sometimes for thousands of miles behind. Nothing could impress upon us more clearly the complex interaction of beach and continent than a tour of the coast from Maine to Alaska. On such a tour we would see more kinds of beaches than most men dream of, some of them stranger than most dreams.

The Northeast Coast

The coastline of New England is too new to have made its peace with the sea. One hundred thousand years ago the rocky islands and cliffs from Maine to Rhode Island were inland hills and mountains. There was no Cape Cod. All along the eastern part of the continent the great Appalachian Mountains, pushed skyward by the buckling of the earth's crust some 250 million years ago, looked seaward across a hundred miles or more of rolling, wooded coastal plain and broad sandy beaches. This coastal plain was and still is an apron of sediments, often several miles thick, which sloped gently from the foot of the mountains to the shoreline and beyond. The submerged part of the apron, between the shoreline and the abrupt underwater slope that leads to the oceanic abyss, is called the continental shelf. When New England had a coastal plain, Boston, Bar Harbor, Portland, and Mount Desert Isle were far inland and the rich fishing areas of Georges Bank and Newfoundland's Grand Bank were fertile coastal fields. The seafloor of the Gulf of Maine is still creased by the course of old streams and rivers.

The shaping of New England is actually a story of many glaciers advancing and retreating for several million years. Coastal New England has been buried and submerged many times, but we will begin our story with its most recent glaciers.

As the last great ice sheets began to gouge their way south some fifty to seventy thousand years ago, they began to form the New England coast that we know. So much precipitation was frozen into the glaciers that they once contained up to ten million cubic miles of water. The world's oceans shrank, and the aprons of the continents extended seaward, the continental shelf rising out of the water as coastal plain. Plymouth Rock was some two hundred feet above sea level and far inland. In northern New England, however, the sheer weight of the glaciers pushed the earth's crust downward, scraping and gouging its surface at the same time. Then some nineteen thousand years ago the climate warmed, the glaciers began to melt and sea level began to rise. The released land of northern New England and Canada began to rebound. Water, as always, was more agile and responsive than rock and earth, and the oceans filled faster than the land rose. The Atlantic flowed inland toward the depressed hills. The measure of the rebound can be roughly gauged by the fact that at the Canadian border clays with marine shells a few thousand years old are found three to four hundred feet above today's sea level. South of Cape Cod, where ice was thin, the depression was negligible.

Most of Maine and New Hampshire might have been flooded if the crust had not begun rising more rapidly between ten and twelve thousand years ago, balancing out the rise of sea level. Meanwhile in southern New England, where there was no rebound, the sea flowed steadily inland. In the last three thousand years sea level has overtaken rebound again, even though the northern coast is still several hundred feet below its greatest pre-ice age height. Like the European explorers, today's captains, vying for the America's Cup, see New England's rocky coast while sailing over an old and gentler coastal plain.

The retreating icecap did not reveal the old forested preglacial landscape. The country that emerged was a stripped, hard land with little soil. The rocky hardship of New England farming is the hardship of farming a land only recently reborn. The image of an even younger New England exists in Newfoundland, where trees still do not grow on vast stretches and where gardeners consider stones the size of a human head endurable.

The jagged nature of the New England coast and its hundreds of small rocky islands are also marks of youth. Travel writers who are knowledgeable about history sometimes wrote of the rock "defying" the waves. The waves, however, are not to be defied. The rock is crumbling. As anyone knows who has marveled at smoothed bits of broken glass and gently rounded cobbles, the sea has a tendency to smooth and round what it touches. What is true of these small pieces is true of the whole New England coast.

Day after day the sea is rocking the giant boulders, throwing cobbles against other cobbles and against the cliffs. Small lichens cling to the rock, secreting chemicals which dissolve the hard minerals. In the long winters water freezes and expands in cracks and crevices, wedging off its bits and pieces. Perhaps the most effective attacks on the cliffs and boulders are the "air hammers" created by the waves. Powerful waves breaking against the rocks compress air into cracks until the air rebounds with explosive force, enlarging cracks and crevices. Eventually all this wearing down will make sand, but the work is almost as slow as the work of the glaciers.

Long, sandy beaches are rare in New England and will remain rare for thousands of years until the sea wears down the exposed rock. The sandy pocket beaches nestled between headlands and the few sandy islands and spits exist because they feed off special glacial deposits left as if meant to provide some token payment for the drowned beaches of an earlier era.

When the glaciers advanced, many little clay knolls protruded into the bottom of the ice. Sand and gravel carried by the glaciers stuck to these knolls. Thousands of these "drumlins" exist in eastern Massachusetts. Some are high, dry, and famous, like Bunker Hill; others, small islands; and still others, submerged shoals. The islands and shoals have eroded rapidly, although on some drumlin islands unsuspecting nature lovers built houses before the waves took their toll. Where homeowners have tried to stop erosion, they have unwittingly cut off the sand and gravel supply for New England's few sandy beaches.

While much of New England's coastline is a "sunken" coast that has never rebounded completely from the weight of the icecap, the famous flexed arm of Cape Cod is a coast raised by the glacier. When a warming climate stopped the advancing ice sheets, billions of tons of rubble scraped from the face of New England lay piled hundreds of feet high in front of the ice. As recently as seven thousand years ago, one massive pile of rubble extended from Martha's Vineyard to Nantucket and an additional twenty-five miles east into the Atlantic. The fishing shoals known as Georges Bank, some one hundred miles east of Cape Cod, were an island, and the familiar form of the Cape itself did not exist.

Created by the glacier, Cape Cod will be destroyed by the glacier's melting. Rising sea level has covered the link between Martha's Vineyard and Nantucket and the island of Georges Bank, which once protected Cape Cod from incoming waves. The outer beaches near Truro began to erode, their material pushed north where it began to form the sandy Provincetown hook. The erosion of the Cape is the normal geological process as the sea takes new material deposited on the coastal apron and distributes it in ribbons along the beaches. Now exposed to the force of the ocean waves and winds, the Cape is a stormy stretch of coast. In a 75-year period the U. S. Weather Bureau recorded 160 gales with continuous winds of over 32 miles per hour. Eight-foot-high waves frequently attack the cliffs along the outer beach and have even overtopped the dunes near Truro, threatening to connect the ocean to Cape Cod Bay through the Pamet River and turn the upper Cape into Provincetown Island.

Erosion of the ocean beaches ranges from one to eight feet per year, according to the U. S. Army Corps of Engineers, which places the entire stretch on the "critical" list. The beaches along the bay side are also eroding, though more slowly. The geological structure of the Cape contains enough material to maintain these beaches for thousands of years. This does not satisfy humans, who want all land to be permanent and all boundaries fixed. While nature goes its own way, a mournful plaque reads: "At this point the sea has carved its way inland more than 170 feet since Marconi built his wireless station in 1902." The antennas and Marconi's shack, which sent the first radio message across the Atlantic, have long since been swallowed by the ocean. The loss of the cliffs has been to a good cause.

The bluffs of Cape Cod, Rhode Island, and Long Island, like the drumlins, are glacial rubble and supermarkets of beach material of all sizes for all weather. Though these stacks of rubble provide wonderful vistas for expensive homes and tourist facilities, the sandy beaches of the Northeast will survive only if the bluffs continue to fall. Indeed, a few thousand feet off the Cape the Atlantic's waves often rise twenty feet high. The reason these giants do not reach the bluffs is because the Cape is protected by a nearshore and offshore bar, both of which have been formed by material ripped from the Cape itself. While many residents and motel owners on Cape Cod bemoan the sea's "attack" on their property, that very attack and its eroded material is protecting them from genuine disaster.

In 1978 the violent February storms sharpened the dilemma of erosion and threatened to stir up a battle among Boston suburbs and towns along the Cape's bay front. In early February, north Atlantic waves rolled in on the mainland, attacking Manomet Bluffs south of Plymouth. The glacial debris, held together by the roots of bayberry and scrub pine, were rimmed with houses built to enjoy the view. The residents have protected their exclusive area with a private access road. When they saw the storm waves eating away the cliffs, the private view became frightening. Their stairs to the beach and their decks collapsed. The owners of the private enclave are now asking the general public to save them. Though homeowners aren't saying so, their pleas for public handouts are probably summed up in the words of the director of the Manomet Bird Observatory (now within fifteen feet of the cliff): "We've decided it's just not worth it to rebuild it if we have to use our own money." The cost of erecting massive stone embankments to shore up the cliffs runs about $50,000 for a sixty-foot lot. The only other alternatives are for the residents to move or to ride the houses down the cliffs. Neither the town of Plymouth nor the Corps of Engineers has been willing or legally able to help. If that changes, or if the residents build their own defenses, the protest will be heard farther south. Lester Smith, a geologist with the state's Coastal Zone Management Agency, has sounded the alarm. Smith notes that the sand torn from the Manomet Bluffs will soon replenish the damaged beaches at Sandwich and Barnstable on the north shore of Cape Cod. If the Manomet Bluffs are locked up behind stone revetments, erosion would accelerate noticeably along the Cape's historic shoreline. Although for thousands of years Manomet and other areas have naturally eroded to the benefit of more southerly beaches, the residents on the bluffs do not acknowledge responsibility to any other towns "downstream." Nor does the legal system in Massachusetts or elsewhere. The laws of private property only recently recognized that the air and water on one person's land cannot be owned exclusively. To reflect the natural facts of life on the coast, property law would have to acknowledge that the substance of the land itself—sand, rock, and gravel—must be shared, or else one person's salvation and pleasure will be another's grief.

The Sandy Beaches

The Gold Coast of Fitzgerald's *Great Gatsby* is the hilly north shore of Long Island with its high bluffs, stony beaches, and intimate coves and harbors. The mansions of the rich in many places still look north across the flooded glacial river valley of Long Island Sound toward New England. When a native of the north shore first travels south across the eighteen-mile width of Long Island to the ocean, he is stunned by the great dunes and in front of them the broad, bright beaches stretching in a straight line as far as he can see. The green waves roll in cleaner and saltier than the water that eddies in and out of Long Island Sound. The beach is smooth and hard and makes the north shore's most carefully groomed public beaches seem like gravel dumps. Though part of the same island, the open ocean beach seems an entirely different world, and indeed it is.

In the eighteen miles from the north to the south shore of Long Island, you pass from the glacial coast to the beginning of the relatively straight, low coast whose sandy beaches stretch, with small exceptions, all the way from Long Island to Texas and beyond into Mexico. This is also the barrier island coast, buffered by narrow islands, each with its own character and citizenry of plants and animals, yet each island vitally dependent on its neighbors. Nowhere in the world is there a barrier island coast either as long or as varied as North America's. To understand this vast whole we must begin with its smallest pieces—grains of sand.

The East Coast from New York to Florida is one of the nation's most heavily traveled landscapes. Its farms and cities are so much a part of our history that hardly anyone realizes that most of the land east of Philadelphia, Washington, Richmond, Raleigh, and Augusta was once not landscape but seascape. This is also true for all of Florida, Mississippi, Louisiana, and great parts of Alabama and southeastern Texas. As sea level rose and fell in counterpoint to several million years of ice ages, the shoreline crossed back and forth across the rock and earth washed from the inland hills. Like a slow broom its strokes smoothed and leveled the great apron of sediment that stretches from Piedmont fall line to oceanic abyss. The only lands flatter than the eastern coastal plain are the salt flats of Nevada and Utah and parts of the Great Plains. These are former bottoms of inland seas. Vast numbers of humans now grow tobacco and soybeans in the sandy soils, and mine the phosphates and limestone left by ancient sea animals. Secure in our travel, trade, and urban cares, we are as little aware of the land's former life as is an empire of ants secure in the warm hollow of an old oak.

The broom that sweeps and levels the coastal apron is a narrow one, an area of high wave energy known as the shoreface, which extends from the coastal dunes seaward for some three to six miles and sometimes reaches a depth of sixty feet. This constant sweeping has kept sand in motion, building both beaches and underwater bars. As a result, the sand deposits on the continental shelf are relatively thin, often less than ten feet deep, and outcrops of rock and old coral reefs are common.

Behind the beaches the smoothness of the coastal plain is interrupted only by rivers and by streaks of low ridges running parallel to the coast but well inland. These ridges are part of an ancient seascape, the remains of old beaches and barrier islands formed between glaciers and left to be colonized by terrestrial plants and animals when sea level dropped. As the shoreline followed the sea, rivers cut through the coastal plain and carried sand for new beaches. These rivers are still delivering the old materials to the modern coast, but not to the beaches and shoreface. The Delaware, Susquehanna, Savannah, Cape Fear, Altamaha, and other rivers yield their materials slowly, and their sluggish currents do not have the great delivery power of western rivers, plunging out of the mountains close to the

Pacific. Most of the sediment that rides these eastern rivers settles into the estuaries and lagoons behind the barrier islands. One day, as sea level rises, this sand may again form beaches. Meanwhile, the beaches are sustained from the more distant past.

A walk on a mid-Atlantic beach is a walk on sand more ancient than the bluffs of New England. These beaches are sustained by reservoirs few people ever see. Most of the sand from the ancient rivers lies offshore in a thin layer on the continental shelf. How does this sand return to the continent? Even at the edge of the continental shelf, at depths of three hundred feet and more, wave action nudges a few grains of sand toward land. The closer they get to the shoreface, the stronger the nudges become until the sand enters the surf. But even the sands from the dark ocean floor once lay far inland. We can prove the inland origin of beaches by analyzing the mineral content of their sand. Light-colored particles of sand are usually quartz and feldspar, minerals so common they might have come from hundreds of sources. The less common dark grains, however, are usually unstable heavy minerals from the original source. Heavy minerals on Atlantic beaches tell us their sands originated not in the nearby coastal plain but in the far inland Piedmont, that rolling landscape of worn-down mountains that rises above the famous "fall line," whose falling rivers attracted colonial mills and industry.

While the heavy minerals give us the specific origins of sand, the general appearance is dominated by quartz and feldspar, making a thousand miles of sandy beaches look all the same. The explanation is fairly simple. With the exception of these two minerals, all others are present only in minor quantities. Quartz and feldspar are chemically stable and physically hard. After millennia of weathering and journeying, only these two remain out of a great variety of rocks. Quartz and feldspar are so hard that tumbling by rivers and tossing by waves do little even to round the sharp grains.

From this longest sandy coastline in the world, the nation demands almost a billion tons of sand a year for highways, homes, offices, pipes, bridges, and ball parks. Sand has become a scarce resource, but not because we can't find it. We simply have run out of deposits that can be mined without digging unfillable holes in valuable real estate. Nevertheless, stripping sand from the Virgin Islands has left those resorts with beaches that are shadows of their natural state. The sand on Puerto Rico's white beaches is being trucked off in sight of signs saying: DON'T REMOVE SAND FROM BEACHES. Where sand mining does not conflict with property values, it threatens the natural environment. Many bays and estuaries, including Jamaica Bay, which was dredged to build Kennedy Airport, have been devastated in the quest for the sand lying in their quiet waters. As our appreciation of wetlands grows, developers look beneath deeper waters.

The U. S. Army Corps of Engineers estimates the offshore East Coast sand supplies at some 450 billion tons, and one recent Corps study estimates 270 billion tons of this sand lies on the inner continental shelf off New York City. Construction interests have begun to eye these deposits for both onshore construction and for the building of deepwater ports and offshore nuclear power plants. The Corps also suggests these deposits might be pumped up onto eroding beaches. The quantity of sand mined may not be as important as the holes it leaves in the seafloor. Because of the costs, "offshore mining" will be close to being onshore. Its holes will change the direction and intensity of waves approaching the beaches. Recent history records one Connecticut beach whose valuable sand disappeared from "natural causes" after dredges mined sand offshore.

While an ancient landscape provided the sand for today's beaches, ancient animals also paid their dues to nature's system, which demands that all parts give as well as take. As shell collectors know, the farther south we go, the easier it is to find beaches strewn with conches, clams, mussels, and scallops. To the "shellers'" dismay, the shells are usually broken. This is also part of the process by which nature maintains the beaches. North of Virginia the shell, or calcium carbonate, part of sand is usually 1 per cent or less. Where warm southern waters come shoreward off North Carolina, the portion ranges from 5 per cent to 25 per cent. And further south, marine life is so abundant that bone and shell fragments are dominant sources of beach material. Not only shellfish but corals, fish, and even microscopic plankton contribute by their death to the East Coast's most famous beaches and most valuable real estate south of the Mason-Dixon Line. In Miami tourists play on a beach that is 50 per cent dead animals. In extreme southern Florida, in the Keys, and in Puerto Rico the beaches are often 100 per cent calcium carbonate. The startling white of those beaches shines from the bleached remains of organisms washed up from nearby reefs. Were the seas not so productive of animal life, those shorelines most likely would be far inland of where they are today, and southern Florida might not exist at all.

The animal origin of the white southern beaches does not make their sands a renewable resource, not in the time scheme of human development. An observant sheller walking the ocean beach should know that the contributing animals are not all modern. Live shells like coquina appear in the surf, but what accounts for oyster shells on the beach when oysters live only in quiet estuaries and bays? Quantity rules out the remnants of old oyster roasts. Besides, the oysters are mixed with lagoon clams and salt marsh snails. Clearly when sea level was lower, the bays and estuaries were in front of today's beaches. Not only do we find ocean beaches made of shells of quiet water species, but of species that have not lived in a given area for thousands of years. The shell content of beaches is largely fossil shell. Three samples of Miami Beach sand, radiocarbon-dated by Dr. Gene Rusnak of the U. S. Geological Survey, showed an average age of some thirteen thousand years, with some material over thirty thousand years old. While development hastens shoreline erosion, its related pollution may be endangering the reefs and animals that should become tomorrow's beaches.

Trees Against the Sea

Everywhere the sea meets land it beats its way forward until it meets rock in some form—cliffs and boulders or the crumbled rock we call sand. From Maine to Mexico there are only two exceptions—the mangrove coast of southern Florida, and the silty "chenier" beaches of Louisiana and eastern Texas. Before we pay homage to the heroic struggles of the mangrove trees, we must recognize the rock on which they stand.

Florida was born out of the sea like Venus. During times of higher sea level, corals and other sea animals formed the limestone that is Florida's bedrock. The luxuriant green Florida we see today, and its sprawling cities, are but a thin film on the ancient seabeds.

The Florida Keys, which are two hundred miles south and west around the tip of Florida, are the tops of dead coral reefs formed some ninety thousand years ago between glacial periods, when sea level was higher. Their white beaches are often "pocket beaches" between rocky limestone headlands. Waders are likely to cut their feet and knees on frequent outcrops of coral.

The Keys were wiped clean by a tremendous 1935 hurricane with winds at two hundred miles an hour. Despite this storm and others like it, rebuilding has always been enthusiastic. The East Coast Railroad line to Key West, stopped by the 1935 storm, has become a highway, and in Florida's southernmost city some thirty thousand people live on an island that could easily be topped by hurricane tides and waves. Some apologist for the present overdevelopment of the Keys will surely point out that at this moment, while developers build motels and hotdog stands, nature is developing a substitute set of Florida keys one to three miles offshore. John Pennekamp State Park, the nation's only underwater park, was founded so that tourists might see the new reef growing. No one will swim on the beaches of this new reef ridge, however, until the polar icecaps again march over the state of Maine and sea level falls.

Meanwhile sea level's rise in the last twelve thousand years has filled the shallow depression behind the arc of the Keys, creating a bay that is still seldom more than ten feet deep. Emptying into Florida Bay is one of the world's least understood rivers.

Flying into Miami from the west, the traveler looks down on almost one hundred miles of uninterrupted wilderness. It is a green wilderness, but the sun flashes back from water as often as it shines on cypress, pine, and savannah grasses. Why is it that the islands in this swamp seem to be elongated from north to south? Because they are islands in a stream. Almost all of Florida south of Fort Myers on the Gulf Coast and Palm Beach on the east is a broad shallow river. The river of the Everglades flows imperceptibly south down the gently tilted tip of Florida.

The "mouth" of the river and Florida Bay are sheltered by the Florida Keys, and in the quiet water here and on the lower portion of Florida's west coast, plants are locked in a complex struggle to extend land in the face of rising sea level. The hero of this struggle is the red mangrove.

To survive in the sea the red mangrove allows its seeds to become complete plants before they drop into the water. The little green twig that bobs on the waves is actually a complete trunk tube. At one end is the soft green bud of the first branch, and at the other the sharp point of the waiting root. For days, weeks, or even a month the twig floats. If it is washed onto a sandbar or mudflat and its pointed end becomes embedded, it quickly sends down roots and rights itself. As the trunk grows it sends out long, arching prop roots that reach through the air like flying buttresses and anchor themselves in all directions. They become a net that catches flotsam and jetsam in a small island of shell, coral, driftwood, seaweed, and mud. Although a single tree might hold itself erect on a pillar of peat only six feet tall and not as wide, frequently mangroves are toppled by storms. Layers of peat mixed with marl on the seafloor bear witness to the trees' innumerable defeats.

Where the red mangroves establish themselves other land plants follow. Black mangroves grow behind the red. More trees, grasses, and vines come along with swamp animals. The marshes and swamps of this wilderness resemble the ancient landscape in which most of the world's coal deposits were laid down. If anyone mines coal in southern Florida, however, it will not begin for several million years, long after the sea has won a final battle and covered the mangroves with more limestone. The limestone that covers much American coal tells the story of how the sea covered swamps before their vegetation could decay, then sealed in the rich carbon with the skeletons of sea life.

For the last three thousand years sea level has risen so slowly that the mangroves, tree by tree, have been pushing seaward. But they now face another enemy in addition to sea level. During the past ten or fifteen years a tiny armadillo-like crustacean called sphaeroma has threatened to do in a few years what rising sea level would have taken thousands of years to do. Sphaeroma bores into the bracing roots of red mangroves and hollows them out within a few days. The mangroves of the seaward edge of the Everglades National Park have almost been wiped out, and the infestation has spread up the west coast of Florida to Naples. If the mangroves disappear and sea level continues to rise, the Gulf will creep in on the southern end of Florida. The Everglades will become seafloor, and the famous "Alligator Alley" highway and the Tamiami Trail will be fragile ribbons of land in a new sea.

The Mating of River and Sea

As if wedded to sand, development of the Florida coast only starts where the mangroves and peat swamps give out at Cape Romano. At Marco Island near Naples the giant Deltona Corporation has been fighting environmentalists and the U. S. Army Corps of Engineers to gain permission to continue filling in over two thousand acres of island swamps so that it might finish a community of some thirty-five thousand people begun in 1964. Just up the coast at Sanibel Island, where Rachel Carson wrote *Silent Spring*, so many vacationers lust after washed-up shells that people get up at night and search the beaches with flashlights.

A few miles north of Tampa in the sheltered northeast corner of the Gulf, an unbelievably flat sea laps a coast without trace of barrier islands. Yet it is here that rivers from out of state begin dumping their sediments. These sediments move south to form the barrier islands from

Cape Romano to Tampa, but where they enter the Gulf wave energy is so low that beach materials move toward shore only during powerful storms. The shallowest offshore profile in the nation is found in the bend of the Florida panhandle. A mile offshore a bather would have to stand on his head to drown. Immediately west of the bend the Apalachicola River dramatizes the special dependence of the Gulf Coast on inland sediment. The river does not carry the clear water from Florida's limestone and peat-bottomed swamps. It originates in Georgia at the union of the Flint and Chattahoochee rivers, and from them it takes tons of red Georgia clay and soil. The red-brown fan of new materials colors the Gulf three miles from shore. The river's delivery power is seen in the sandy beaches and the long straight barrier islands formed from its burden.

Further west the Mississippi River is actually pushing the claims of the land seaward. Every day of the year it delivers over a million tons of sediment. Every seven to nine thousand years it has transported a one-foot-deep layer of soil from its entire watershed, which extends north to Minnesota, east to Ohio and North Carolina, and west to Montana. As the delta grows it sometimes encloses part of the Gulf, leaving pieces of the sea behind as lakes. Lakes Borgne and Pontchartrain near New Orleans are both imprisoned parts of the Gulf. But for its own weight, the Mississippi delta, already some twelve thousand square miles, might have filled an even greater expanse of the Gulf.

The visible delta, carpeted with marshes and cotton plantations, is only the surface of a gigantic pile sometimes several miles thick. It is pushing down the rock beneath it and hastening the rise of sea level to the east in Mississippi and to the west in Louisiana. But the delta, despite its weight, continues to grow on top of itself. The river carries so much mud that it clogs its own channels and is constantly seeking new paths to the sea. The mudflats created by the river's many channels are quickly turned into land by plants. And for almost a hundred years now plants have been invading the channels themselves. The Brazilian water hyacinth grows so fast and so completely covers the surface of a sluggish bayou that it shades out almost all life beneath its leaves.

The delta coast is a marshy prairie and most of its beaches are a dark mixture of sand and mud. Its first regular visitors were pirates like Jean Lafitte. As if waiting for the land to firm up, towns and cities have generally stayed away from this new edge of the continent. Only thousands of oil towers speckle the dark offshore waters, pumping out the riches from an earlier coastline. The relatively fleeting life of an oilwell and the rewards that match most risks make the delta acceptable to drillers. On West Timbalier Island, since 1964, Gulf Oil has spent more money to build rock walls in front of its production and storage facilities than the island itself is worth. The island, which has migrated more than a mile since 1907, has now lost most of its beaches. The local Corps of Engineers recently rejected pleas for more defensive rock work, and their local engineering expert predicts that with the next six-foot-storm surge the island will be "lost." Actually the island will jump backward, leaving oil tanks and their defenses in the Gulf of Mexico.

Those who cannot afford to be abandoned to the sea must choose their islands carefully near the delta. Most important, you must know which delta you're near. There is not one but many Mississippi deltas. Scientists using soil and seafloor borings dated by radioactive carbon content have located seven deltas formed in the last five thousand years. Only the modern delta, where the river now enters, is pushing seaward. To the east and west, where the river's old channels long ago choked on mud, the shore is retreating and the river's work is being erased by the Gulf. The final effect (at least since maps were first made) is that Louisiana has lost more land than it has gained. Because the delta's materials are so silty and light, erosion rates up to fifty feet a year change landscapes beyond recognition in less than one human generation. Many delta residents have found out that a single storm may destroy whole islands and produce a lifetime's changes overnight. Average erosion rates are of more use to statisticians than to owners of shorefront property.

West of the protruding delta the kingdom of the Mississippi River extends some three thousand miles into Texas. This shoreline is smooth, and behind it lies a vast marshy prairie. It is a land newer than New England and built on the mud and clay that started drifting along the shore when the Mississippi River entered the Gulf west of its present mouth. As the Egyptian pharaohs built their pyramids, the river's sediment created a great plain in Louisiana and some of the world's richest marshlands.

The new land, unlike New England's rocky shore, was extremely vulnerable, and its troubled history shows in a flat plain with a peculiar wrinkled face. The sea was its creator and always its master. Whenever the river mouth shifts back to the east, where it is now, the material part of the dynamic equilibrium yields to energy and the Gulf erodes the fragile coast. Waves drive the beaches landward, but the same waves sort shells, gravel, sand, and driftwood from the silt. Gradually this material is pushed into a ridge which, when solid and high enough, slows the sea's advance. Three to ten feet high and as much as six hundred feet wide, scores of these ridges, each formed during a period of erosion, streak the marsh prairie. On top of them grow the scrub oaks (*chênes* in French) which give these ridges their name—cheniers.

Near Holly Beach the only coastal highway in the chenier plain, Route 82, washed away regularly in storms even after the U. S. Army Corps of Engineers had spent several million dollars constructing a four-mile-long rock wall, or revetment, along the beach. A cheaper solution would have been to move the road back onto the higher ground of a chenier ridge. Sources in the Corps say this was vetoed by the political quarterbacking of a prominent New Orleans businessman-sportsman.

The retreat of the western Louisiana shoreline will continue as long as sea level rises and the Mississippi continues to empty at its present location. The Corps of Engineers, who have not yet attempted to tinker with sea level, have however kept the Mississippi in the channel it began some six hundred years ago. For a while the river threatened to push over into the channel of the nearby Atchafalaya River and start a new delta far west of its present mouth. This might have meant the eventual growth of many square miles of western Louisiana coastline (and erosion in the east). The Corps and a company of cranes and bulldozers and dredges have kept the breach under control, allowing the Atchafalaya only a small theft of water and silt. It has been enough, however, for some coastal building just east of the smaller river's mouth. This puts the Corps in the eventual position of directing a lot of beach building in southern Louisiana.

The great losses of soil from the American Heartland have been Louisiana's gain. Its geologically new land contains over five million acres of marsh. Many birds that breed and summer in the upper Mississippi and Missouri basins unerringly find their way down the great midwestern flyway to winter in Louisiana marshes many times richer and more productive than the finest Iowa cornland. The creole shrimp enjoyed by diners in New Orleans also come from these marshes.

In western Louisiana small sandy beaches begin to appear as the silts of the Mississippi thin out. The beach becomes firmer underfoot, but in most places no one spreads a beach blanket in the oily mud. At the Texas border where the Sabine River enters the Gulf, swamps give way to grassy marshlands, and a fringe of pink and white sand meets a heavier sea. The fringe extends eight miles to Galveston, where suddenly the beaches are broad and bright. We have returned not only to sandy beaches, but to the barrier island coast.

Though Texans deserve none of the credit, the barrier islands here are the longest and broadest in the United States. On most Atlantic barriers you can pull a boat up on the bay side and walk less than a mile to the ocean. In Texas the walk could be a hot, dry five miles. The union of Mustang and Padre islands in south Texas, joined except during storm tides, forms a single beach 130 miles long.

These record large islands were once considered wastelands. Francis Sheridan, a British diplomat, sailed into Galveston in 1840 and later wrote: "The appearance of Galveston from the harbor is singularly dreary . . . It looks like a piece of prairie that had quarreled with the mainland and dissolved the partnership." That was back in the days when Galveston itself was a burly little port of cotton shippers and fast-buck traders. Today, with both Galveston and Houston grown rich, the sands are coveted by vacationers, developers, and land speculators. Some twenty-five miles, at the north and south ends of Padre Island, are gaudy with commercial development, its tiny beach lots being advertised nationwide, as when the Florida coast was being sold to Depression-reared Yankees. In between these two growing beach cities lies eighty miles of National Seashore. Motor vehicles are banned and the entire stretch remains almost as quiet and clean as when Spaniards, in galleons, stared across the island wondering if the cities of gold lay inland and not knowing they heard the surf breaking over the two resources Texans treasure most—oil and sandy beaches.

Unlike the Atlantic islands, the islands of Texas are being nourished directly by rivers. Though Texas is a semiarid state, its coast is fed by many rivers whose channels wind hundreds of miles north and west toward Dallas, Abilene, Waco, and across the entire state into Colorado and New Mexico. The sand budget for the islands, like so many other Texas budgets, is ample and continuous, and the islands grow long and fat. The Rio Grande and Brazos rivers in past epochs have built deltas into the Gulf. Today human intervention and rising sea level have eroded the old deltas. Irrigation and dams have halted the flow of sediment. The Gulf can now distribute to the barrier islands all the sand the rivers bring.

Southern California: Don't Build on Fifth Street

No one living today remembers Sixth Street in the town of Encinitas north of San Diego, yet a subdivision map filed in 1882 shows it clearly. The developer expected Easterners to come and buy prime lots where they could build between the road and the edge of the cliffs overlooking the Pacific Ocean. Some residents now perched between

Fourth Street and the cliffs say the map is fiction. A few longtime residents, however, remember a Fifth Street. In 1946 officials closed it. Today it has disappeared.

All along the California coast, but most spectacularly in the south, the ocean is reclaiming the land. Not claiming but *reclaiming*, because the materials of these cliffs were underwater once before. A higher sea level between glaciers tells only a minor part of the story. The other part is told by the cause of California's earthquakes. As one great plate of the earth's crust meets another, the southern California coast is bending upward. While sea level may gain on the cliffs between episodes of bending, over the past forty to fifty million years the coast has been rising. If the mind's eye could see history as a film unreeling at the rate of several thousand years a second, we would see this hilly coast rising jerkily out of the ocean. For brief seconds the old ocean floor would lie exposed to the air. Then the sea would cut its way forward, creating a new cliff in the sloping terrain. In many places the coast falls in a series of terraces, each terrace an old seafloor. The coast which resulted from this hesitant emerging of seafloor is built by layers of hardened sediment, with here and there some lava from an ancient volcano.

Fifth and Sixth streets in Encinitas did not disappear because the ground on which they sat was ancient, but because it was seafloor. Most of the coastal hills and cliffs of California are made of sandstones, shales, and siltstones. The durability of these rocks varies, but in most cases, at least near cliffs, developers should heed the biblical warning about houses built on sand. Under the assault of the ocean these sedimentary cliffs are more like loose sand than rock.

Streets were not the only thing to disappear in Encinitas. In 1938, some nine years before the disappearance of Fifth

Street, the Self-Realization Fellowship built what was said to be the largest building on the southern California coast. The four-story-high temple on the bluffs overlooking the ocean cost $40,000 to build in the midst of the Depression and was complemented by two large pools. Leakage from the pools lubricated the loosely layered rock beneath the foundations. Then in the winter of 1941 a heavy rain soaked the ground and large waves cut into the base of the cliff. The proud temple and its towers bowed toward the ocean and pitched forward. These cliff collapses are so sudden and unpredictable that that same winter two children playing on the beach were crushed to death, and further south at Del Mar a freight train on the coastal route suddenly found itself in midair. Cliff erosion had undermined the tracks. Fortunately, a passenger train full of New Year's Day travelers had been delayed, and the freight train's crew of three were the only fatalities.

The train wreck, Fifth Street, and the Self-Realization Fellowship Building are hardly remembered. Even railroad officials do not remember or know of the old track, and geologists had a hard time in 1977 persuading them that erosion was about to dump another train. That year storms removed protective rubble from the base of the cliff, clearing the way for later storms to attack the cliffs themselves. Storms caused miles of cliff failures. When the trains were finally stopped, only seven feet separated the tracks from the cliff and the ocean, and ominous fissures were already opening in this thin margin.

Rubble at the base of a cliff, even if it does not contain bits of houses and railroads (as it often does), should tell any amateur geologist that danger lies above. As soon as one storm clears the rubble, the next could bring down the cliffs. In addition to this obvious clue, geologists note three conditions conducive to landslides: first, landslides often happen when a rainy season lubricates layers of rock; second, shale and other rocks which are bedded in smooth, thin, flat layers are most likely to slip; third, layers which dip toward the beach are most likely to give. The actual mechanism is a little more complicated. Under normal conditions, debris collects at the base of the cliff, shielding it from normal waves. Meanwhile the upper edge of the cliff retreats as drainage erodes it or small sections slump. Sooner or later the face of the cliff reaches an "angle of repose" which is more or less stable until storm waves remove the protective rubble at the base. The cliff then becomes too steep and another cycle begins. Many California cliffs are also riddled with caves, cracks, and rock joints which collapse and accelerate cliff erosion. Ideal conditions for cliff collapse occur when a permeable layer of rock tilts toward the sea and lies on top of an impermeable layer. Water percolating down settles on the impermeable rock and greases the skids. This is the condition at Del Mar just north of San Diego.

Developers of a large cliff-top condominium (called "the Great Wall" by the vice-president of the town council) say fears of disaster are alarmist. Using recent measurements of erosion, they can show that the cliffs are retreating only a half inch per year. The town councilman says the Great Wall has not only walled citizens away from the beach but has walled the developers away from reality. He and geologists at the nearby Scripps Institution of Oceanography say this is a case of how to lie with statistics. They point out that average rates of cliff retreat are meaningless, since the cliffs slump suddenly, in large blocks, and more in one place than another. Developers can support their claim of a half inch a year by taking a large section of coastline and averaging the cliff retreat over the past ten or twenty years when the coast has been free of major storms and big waves. But each locale has its own history and geology. Gerry Kuhn, a La Jolla geologist, used old tax maps, railroad, and land surveys to show up to eight hundred feet of shoreline retreat in Encinitas. Half of this loss followed severe storms and floods between 1883 and 1889. Stories of cliff retreat is reflected in the regular disappearance of certain maps from the official file. Kuhn says the shoreline retreat also shows up in tax assessments. As a lot begins to fall in, its value goes down. When it is no longer a viable cliff-edge lot, the records show a dramatic rise in assessment for the lot behind it. A study done by the Southern California Testing Laboratory for the Del Mar Beach Club development in 1976, Solana Beach, concluded that within no more than three to five years cliff failure, unless stopped, would destroy the clubhouse and a residence.

Even the National Marine Fisheries Building, built in 1963 on the Scripps Institute campus, is disastrously located. The "Tuna Hilton" rests partially on a piece of bluff known as a slump block. Designers say the building is specially articulated so that it should stay intact as the bluff falls from underneath its seaward end. Scientist George Moore, writing in the 1973 guidebook of the San Diego Association of Geologists, says: "Inasmuch as my own office is near the east [landward] corner of the building on the ground floor, I am especially interested in the correctness of this conclusion."

Development itself has helped turn the cliffs into sand. To enhance the ocean view, builders usually remove the small dune lines or ridges from the edge of the cliffs. Water then drains freely over the edge and down the face of the rock. Pipes from storm drains within the development also emerge on the rock face and empty with the force of small waterfalls. Even lawn watering increases the danger as water seeps into the permeable rock layers. Lawn watering is so heavy in dry times that the water level in many local wells actually begins to rise.

Nature will be indifferent, as usual, to human problems. The erosion of cliffs and headlands all along the California coast not only feeds new material to the beaches, but leaves small islands of boulders a short distance offshore. These islands often become important resting places for birds and seals.

If California's collapsing bluffs, like those of Cape Cod, could be the main source for sandy beaches, erosion would be a blessing in disguise. The famous strands of Muscle Beach, Mission Beach, Santa Monica, and Malibu, however, all owe their existence in large part to rivers. Under natural conditions the river basins in the lower 240 miles of California receive some twelve million cubic yards of sediment each year. This load would seem to guarantee endless youth for the beaches. The system is not so simple.

Because the submerged continental shelf is so near the leading edge of the continental plate, the shelf is narrow and sand disappears quickly over its edge into the oceanic abyss. As if this is not problem enough, the southern California shelf is creased by six spectacular submarine canyons, long narrow gorges whose dark fingers reach shoreward and pull masses of sand and silt to depths from which they will never be recovered. Every year the Scripps Submarine Canyon near San Diego swallows up to 200,000 cubic yards of sand, enough to make a beach 150 feet wide, 18 inches deep, and over 4 miles long.

In 1975, Scripps Institution and Cal Tech began a comprehensive study of beach material supplies in southern California. The study reaches back in time to 1925 when Hollywood was in its infancy, when the aerospace industry was little more than science fiction, and when refrigerated trucks and trains were not streaking eastward with California fruits and vegetables. The study is not complete, but already sketches a region strangling in the tentacles of its own desires.

An exploding population has invaded the coastal hills and flood-prone coastal lowlands. People have outstripped the water supply, and their highways, houses, and offices clamor for building material—often sand and gravel. Today the region's construction industry annually mines over twenty-three million tons of sand and gravel from beaches, riverbeds, and dunes. The case of the Santa Clara River north of Los Angeles is a dramatic illustration of how much beach material is locked up by watershed "management" and "development." Between 1928 and 1955, human structures reduced the deliveries to the ocean by only 6 per cent. Since then, three more dams have been built, each taking its toll until the river's delivery capacity has been cut some 37 per cent, which includes an annual loss of sand alone amounting to seventeen million tons. The Santa Clara's "clarification" means that in the past twenty years the region has lost enough sand to create a beach 50 miles long, 110 yards wide, and 1 yard deep.

The Santa Clara's tale is repeated on each of the nine major rivers in the region. The Ventura River has lost 66 per cent of its sediment. Throughout the area beach material is manipulated, locked up, or consumed by 311 water supply lakes and flood-control reservoirs, 77 sand and gravel mines, and 65 percolation basins. With the narrow continental shelf and the submarine canyons creating a constant demand for renewal of beach sand from inland sources, the effect of urban and agricultural growth has been disastrous and perhaps irreversible.

With natural supplies of sand cut off, many towns have turned to the U. S. Army Corps of Engineers, asking them to find new sand for starved beaches. Since World War II the Corps has obliged with more millions of dollars' worth of sand-pumping operations. During the last five or ten years artificial beach nourishment has moved over two million yards of sand a year, almost as much as the natural system once did at considerably lower cost.

California's insistence on building cities and cultivating arid lands has been mirrored in Arizona, and together the most bitter fruits of this unplanned growth has been borne, not in this country, but in Mexico, where the Colorado River enters the Gulf of California. Once the beaches of the upper Gulf were supplied with 180 million tons of sediment a year by the river. California's and Arizona's thirst and greed for hydroelectric power have utterly changed this. In 1935 the United States built the Hoover Dam. Then came Glen Canyon, Parker, Imperial, and Laguna dams. On the tributaries were built Flamingo Gorge, Navajo, and Curicants dams. Today the Colorado River delivers almost no sediment to the Gulf. Irrigation water is returned diminished, salty, and loaded with pesticides. Here, as on Egypt's Nile delta, starved by the Aswan Dam, the lack of arriving sand has allowed the waves to eat into the beaches. The tides run faster and heavier in and out of the once fertile estuary at the river's mouth, and the shoreline has eroded back as much as five miles. Commercial and sports fishing boats find that without nutrients fish populations have declined. In 1969 the shrimping industry folded its nets and disappeared.

A Battered Shore of Cliffs and Coves

The disappearing beaches of southern California, like its fashions and movie trends, may be the image of the future. North to San Francisco and beyond, all the way to Cape

Mendocino in the redwood country, the beaches rely on inland erosion for nourishment. Rivers are the major source of material. Yet growing cities increasingly clamor for dams and reservoirs. This section of California coast is an almost continuous wall of spectacular cliffs and bluffs. Long, broad, sandy beaches seem more prevalent than they are because they are so often boasted about by chambers of commerce—a sign of scarcity.

This is essentially a coast of coves and creeks and cliffs. This is especially true north of San Francisco. A driver on the coastal highway begins to notice that every few miles, with almost mathematical regularity, the road veers inland to descend the sides of a sharp canyon, cross a creek or small river, and ascend to the sea cliffs again. The bluffs bordering the canyon offer aerial views of a typical beach pattern. The ocean, as it receives new sand delivered by the waterways, pushes it into ridges. Waves, generally from the southwest, shape the ridges into north-reaching spits which would completely close the canyon mouths except for the narrowest exits cut by the fresh water. Wave energy is so high and sand supply from this region of dry summers so small that these spits and bars are often the only sandy beaches for many miles.

While the ocean seems to be trying to block the creeks and rivers with their own sand, the little spits provide effective protection from the harsh sea. Behind them lie the region's few coastal marshes and excellent little campgrounds full of driftwood. The complete camper might have a surfboard for riding the ocean waves in front of the spit, and some diving gear for gathering the tasty abalones for the evening meal.

From San Francisco to Eureka, over two hundred miles, the coves are so small that the coast offers only one or two ports, small villages sending out timber from the mountains and harboring the growing fleets of pleasure craft owned by the folks on top of the cliffs. On this high coast, as nowhere else in the country, mankind's grasping after magnificent views is evident and dangerous. Houses crowd the sharp edges, their picture windows looking down at the power of the breaking waves and west across the Pacific to the unbroken horizon. So recently have subdivisions come that as we took pictures of the clifftop houses in Irish Cove Point, we saw a half-dozen deer grazing in the vacant lots next to us. So great is the demand for the cliff view that just south of Point Arena twelve solid miles of coast between highway and ocean is now developed as the exclusive Sea Ranch subdivision. The former sheep ranch is a potential new town of twenty thousand people. Many of the houses on this coast could well become driftwood before the mortgages are canceled. And since the cliffs the houses rest on also lie on or near the San Andreas Fault, cliff erosion could become an academic question any day.

At Cape Mendocino the coast changes rapidly. So does the weather that helps shape the landscape. The long, dry summers give way to the wetter weather typical of the Northwest. The rivers flow amply all year, and when they have not been dammed up, they bring sediment to the coast. On the coast itself the southwesterly winter winds combine with the highest surf in the country to drive sand northward. Milder summer winds from the north return only a portion of the sand. The result is a coast with many north-reaching sand spits and long, straight beaches often backed by dunes where there is space between mountains and sea. Like their smaller images to the south, these sand spits form perpendicular to the mouths of the rivers. Behind them lie quiet bays and harbors.

Although the rivers in this rainy and mountainous area discharge great quantities of water, there are no protruding deltas like those of the Gulf Coast. First, there is much less sediment contributed by the smaller mountainous drainage area than by the flatter, soil-laden Great Plains and by the Piedmont of the East. Second, the intense energy of the Pacific's waves quickly disperses concentrations of sediment.

More sand, more water, and stronger waves have created larger spits, bigger harbors, and broader lagoons and marshes. Eureka, Crescent City, Coos Bay, Newport, and Astoria are all well-used seaports. A significant part of their beaches are the remnants of their principal export—lumber. Stumps, crotches, roots, and limbs, stripped of bark and bleached by salt and sun, lie in a fifty-foot-wide rift along the foot of the southern highway approach to Crescent City. On a warm day the debris tints the air with the sweetness of cedar and redwood. Perhaps there is poetic irony in the remnants of the great trees protecting the roads along which the logs from yet other great trees are trucked into the sawmills, shipping docks, and freight platforms.

Before we follow the great breakers and cliffs to their final Northwest grandeur, we should pause at the lower and more complex scenery of the coastal dunes. Although 90 per cent of the scenic postcards show cliffs and sea spires, almost half of the Oregon and Washington coast is backed by sand dunes.

The largest coastal dunes in North America stretch for almost fifty miles along the central Oregon coast. Heading north from Coos Bay, riders on the coastal highway often think the road has veered behind the hills when, in fact, those hills are tree-covered dunes, some up to two hundred feet high. The age of the dunes is revealed in cross-sections showing soil layers built by thousands of years of vegetation. This great reservoir of sand has a complex origin in fifteen thousand years of rising sea level and constant winds blowing landward from the beaches. Summer waves pushed river-borne sand onto the beaches, and before it could be taken back by winter storms the wind lifted it onto the dunes. As the grasses, brush, and trees grew and trapped more sand, the dunes grew taller. Gradually a rising sea level pushed the dunes inland. Where the dunes are naturally active or where grazing and occasional cottage development have stripped the vegetation, the march of the dunes is unmistakable. Sand pours down the back side into the little fresh-water sloughs and ponds along the highway. Long fingers of sand reach into the evergreen forests, burying the trees beneath new dunes. So long as sea level continues to rise, the landward advance of the dunes is their salvation. Thus, near Coos Bay where moving dunes have been artificially stabilized, sand no longer blows inland and the infant dunes of the future are dying of starvation.

Where smaller dune fields occur on the Oregon coast, homeowners have often built on them or behind them, believing the dunes to be as permanent as the hills farther inland. Such is the case of Siletz spit near Lincoln City. Shortly after building began in the 1960s, residents were stunned by winter storm waves over twenty feet high carving away the dunes. Year after year they have trucked in fill and placed boulder rip-rap along the face of the dunes. Ironically, part of their problem was that, in preparing the lots and selling the houses, builders had removed many of the big logs that waves had strewn everywhere. The importance of the drift logs should have been clear as soon as the dunes began to collapse. Protruding from the sand were yet more logs, as much part of the dunes as two-by-fours are parts of houses.

Not far north of the Oregon Dunes is the greatest single source of sand on the West Coast, the Columbia River.

Unfortunately for eroding beaches in Oregon, most of this sand is lost on the steep continental shelf or carried north to help Washington's beaches. The rugged headlands and the deep embayments of Washington's shoreline prohibit the Columbia's contribution from being spread very far north. For good reason painters and postcard printers picture the Northwest as a coast of rocks, cliffs, and waves. Nowhere else in the country outside Alaska has higher cliffs or a more powerful surf. The farther north we go, the higher the cliffs and the wilder the country. The last 140 miles of Washington coastline is the wildest coast in the lower forty-eight states. From Gray's Harbor north to Cape Flattery on the Canadian border, the beach is backed by cliffs that often rise three hundred feet or more. Only an occasional river breaches this wall. An unusual interaction of climate and shoreline occurs here because of the high rainfall, up to 115 inches a year. Nourished by these rains and a rich soil, great spruce trees grow close to the margin of the sea. The shoreline is littered with their huge trunks. Most of the year the driftwood protects the beaches and cliffs from erosion by holding the sand and breaking the force of the surf. But in the giant surf of winter storms the driftwood fills the sea with battering rams pounding the weak shale and sandstone of the cliffs. In the wet stormy weather landslides are also common, and the material knocked loose quickly becomes next summer's beaches.

The Alaskan Coast—Another Book

A single powerful lobe of the last glacier carved out the straits of San Juan de Fuca, which separate the United States from Canada. The 400-foot-high rock promontory of Cape Flattery, overlooking the straits, signals the beginning of the glacial shoreline. In Alaska glaciers are actively shaping the coast, carving deep fiords, pushing seaward masses of rock and mud large enough to bury cities or to build them on. (As we'll see later, the result may be the same.) From the glacier's lips huge icebergs drop into the sea, with their ripples rolling seaward as tidal waves.

Although the glacial fiords predominate, the Alaskan coast contains everything we have seen on our tour of the lower forty-eight states—cliffs, barrier islands, spits, marshes, and deltas. It also offers a few arctic beaches that occur nowhere else in North America. Excluding bays and fiords, the Alaskan coast is 6,800 miles long. The combined Atlantic, Pacific, and Gulf coasts measure 5,100 miles. Yet to the joy of some and dismay of others, the northern coast will remain unknown and unknowable for several more generations. The Alaskan coast is a book in itself.

With the exception of Alaska, the tour we have just completed may give the impression that we know a great deal about what beaches are made of. Describing materials is almost as easy as saying biscuits are made of crumbs. The materials are there, we collect them, subject them to modern analytical tests, and give them a name. But beach materials can change within the space of a few weeks. We would be tempting ourselves to think of the beaches simply as collections of debris, surfaces as unchanging as the rocks of Central Park, the stone faces of Mount Rushmore, or the swamp prairies of the Everglades. All the world's landscapes change, but few change as fast as the beaches. Beach materials are often here today and gone tomorrow. The materials of the beaches exist on the edge of the continent like sets on a stage. The other three parts of the dynamic equilibrium—sea level, energy, and shape—constantly rearrange the scene. To understand the drama we must turn our attention to these actors who make and possess their own stage.

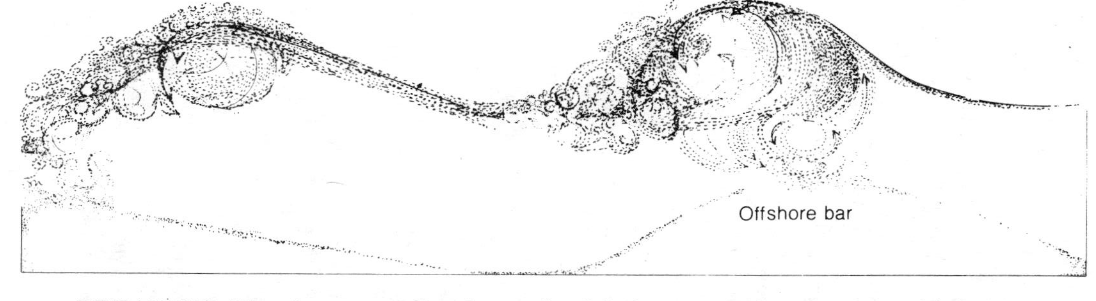

BREAKERS. *The plunger-type breaker to the right has a crest that rises up and falls into its own trough, sometimes compressing an explosive tunnel of air in its curl. The spiller-type breaker to the left breaks over a more gradual beach slope, and its crest tumbles down its own forward slope.*

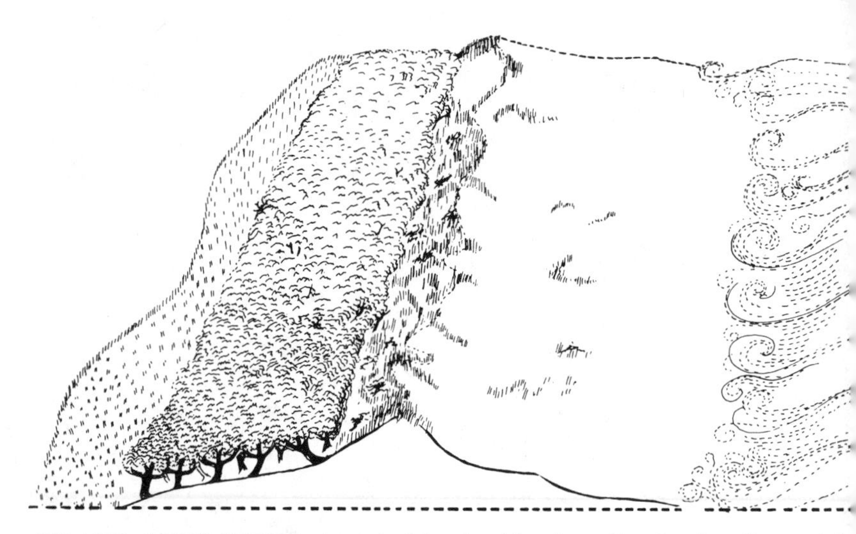

ISLAND MIGRATION. *A typical barrier island profile showing (from righ left) waves, beaches, dunes, maritime forest, and marshes. The initial forest gro thrives in the shelter of the dunes, and the forest's closed canopy protects fur growth from salt spray.*

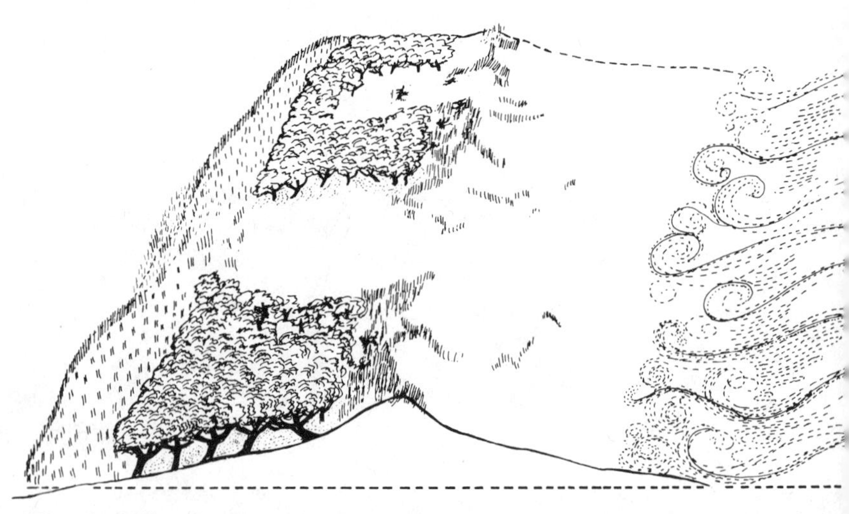

The dune line has been breached. Sand has buried part of the maritime for while other parts of the forest have been killed. The island's elevation is risin these points.

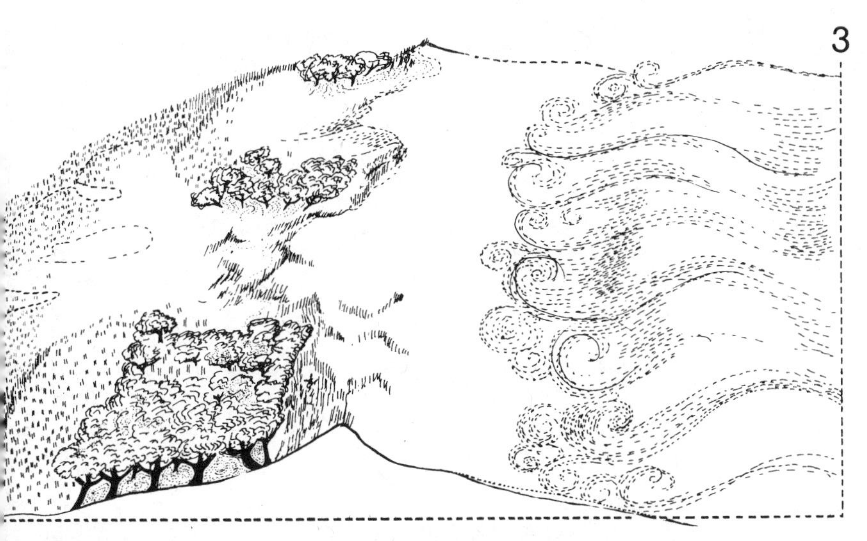

ing sea level and storms have overwashed much of the island, pushing sand all way to the lagoon and thereby widening the marshes. Vegetation is beginning grow in the old overwashed sand where new dunes will form.

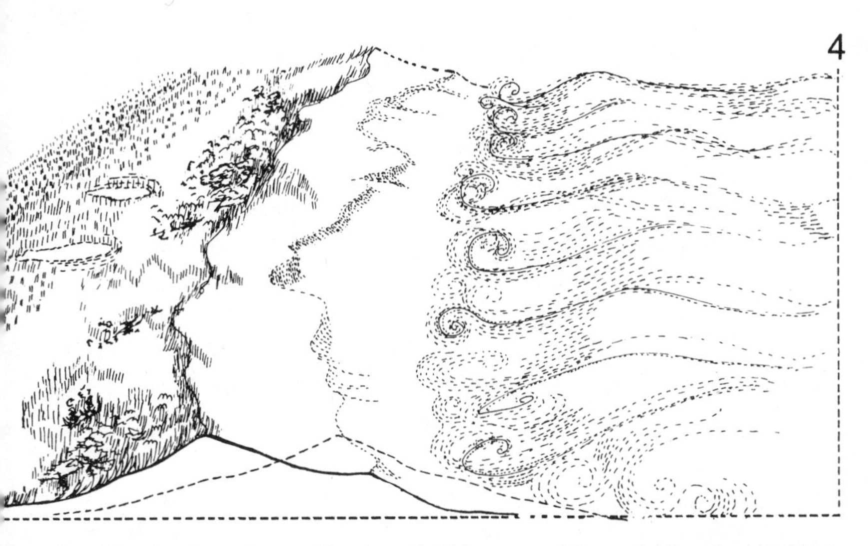

new dune line has been formed by the wind-blown sand from the beaches. Behind e beaches the forest is recovering, and a new marsh has grown to extend beyond e old one. If the sea level rises very slowly, the total volume of sand in the island mains about the same. If the sea level is rising rapidly, the island may become inner as it migrates.

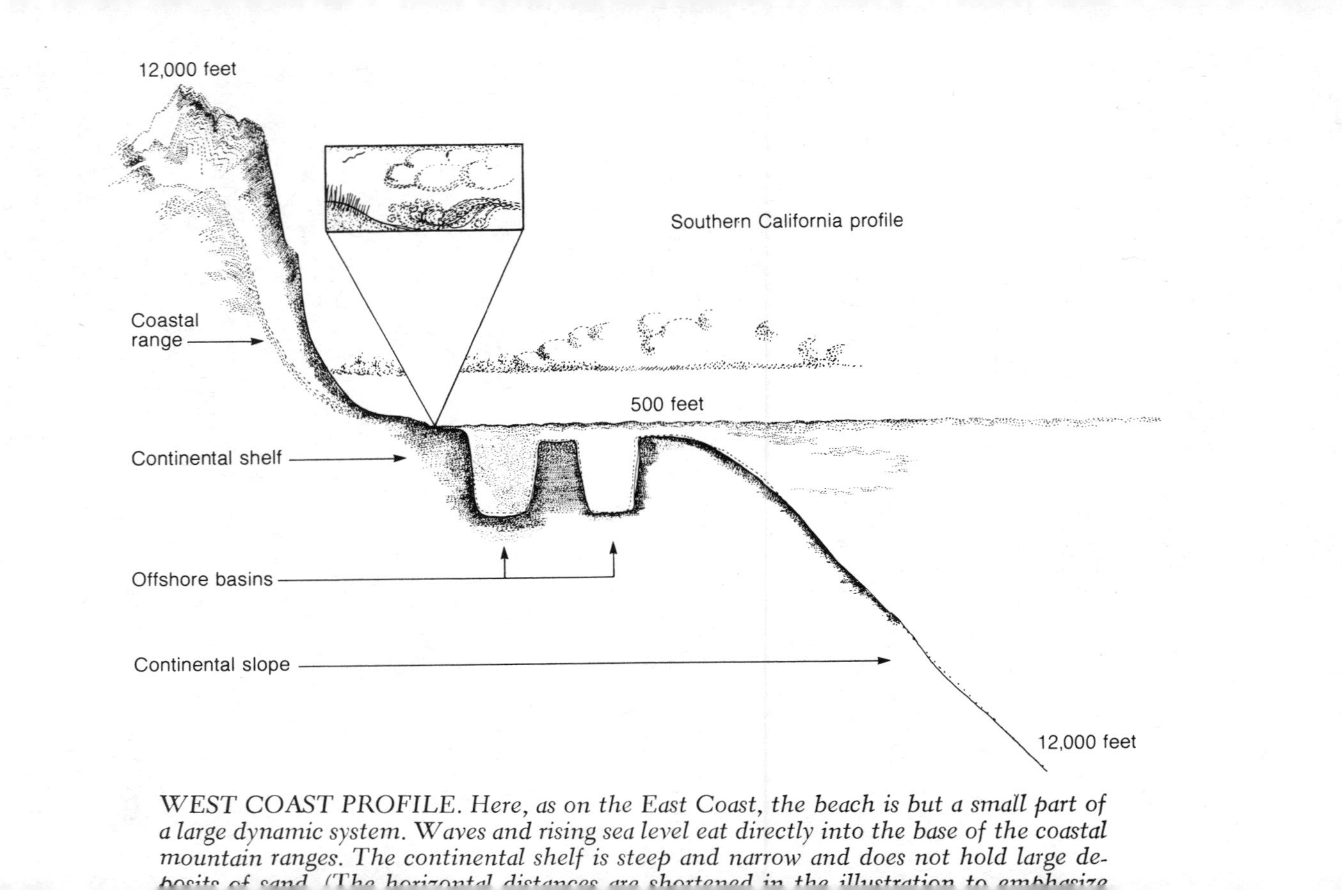

WEST COAST PROFILE. Here, as on the East Coast, the beach is but a small part of a large dynamic system. Waves and rising sea level eat directly into the base of the coastal mountain ranges. The continental shelf is steep and narrow and does not hold large deposits of sand. (The horizontal distances are shortened in the illustration to emphasize

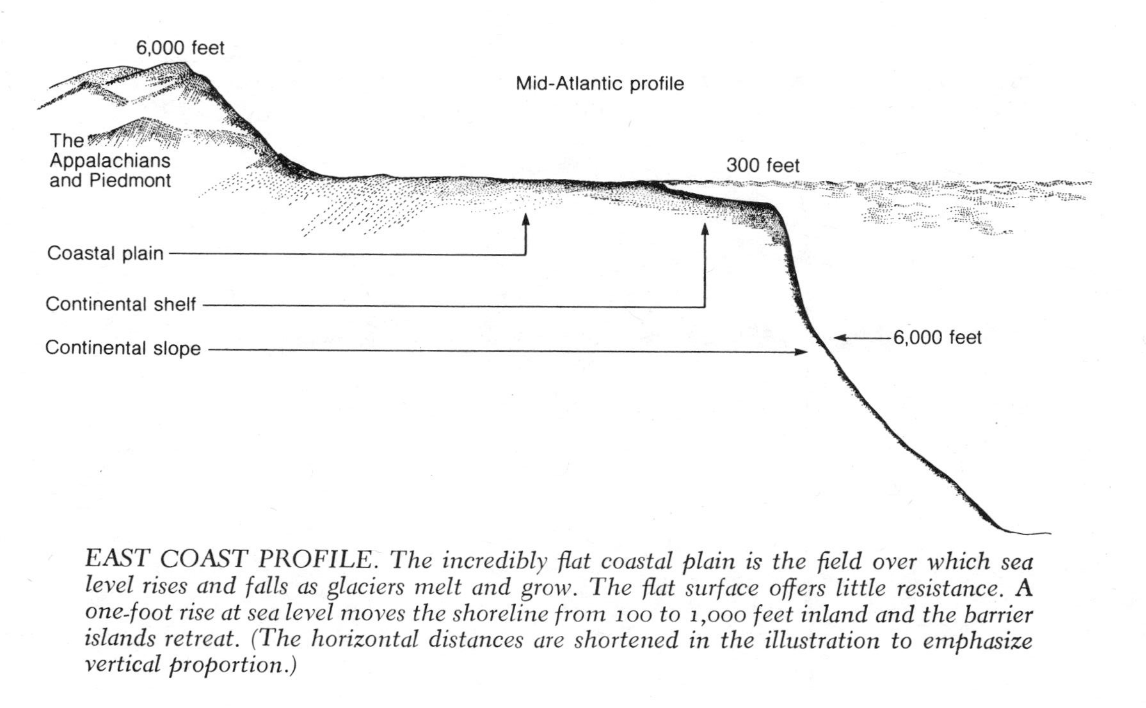

EAST COAST PROFILE. *The incredibly flat coastal plain is the field over which sea level rises and falls as glaciers melt and grow. The flat surface offers little resistance. A one-foot rise at sea level moves the shoreline from 100 to 1,000 feet inland and the barrier islands retreat. (The horizontal distances are shortened in the illustration to emphasize vertical proportion.)*

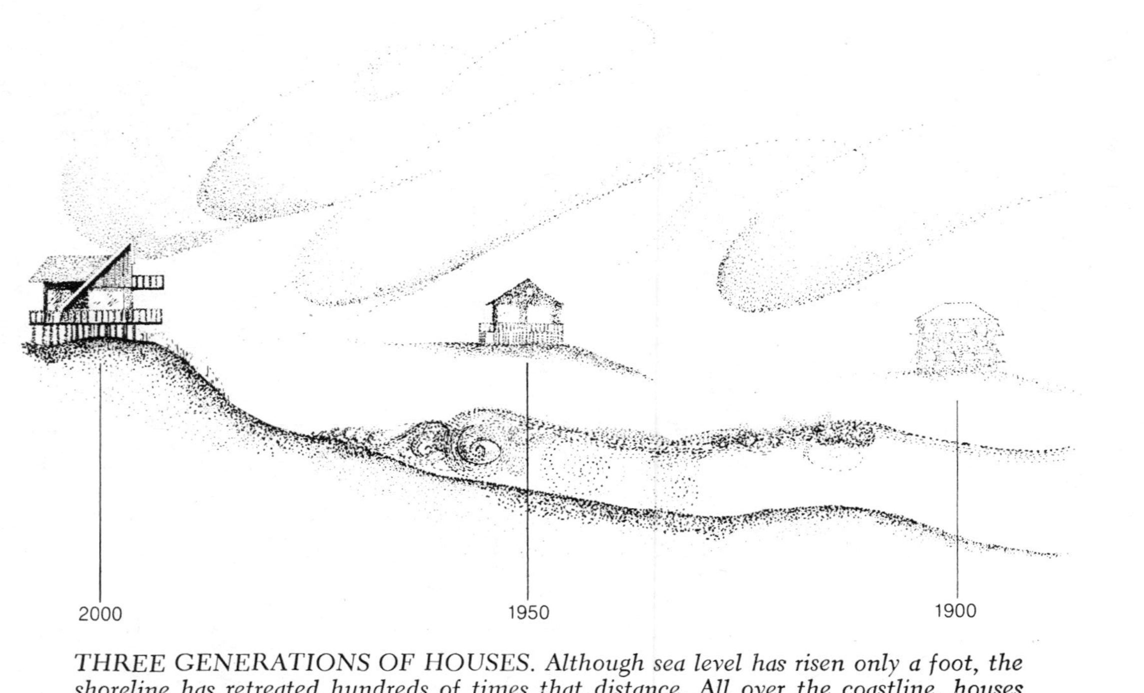

THREE GENERATIONS OF HOUSES. *Although sea level has risen only a foot, the shoreline has retreated hundreds of times that distance. All over the coastline, houses must be moved or fall into the surf.*

4

Energy: Winds, Waves, and Tides

MARINERS HAVE OFTEN called the surface of the sea barren and thought of their ships as tiny isolated travelers in a water wilderness. To the frightened or bored passenger a porpoise swimming along in the bow's wake, or a sea bird winging on its mission, is welcome company. Perhaps the root of our midocean discomfort is the sense of being without direction, without points of reference. Except on a dead-calm sea, however, the trained eye sees direction everywhere in the waves.

Waves always have direction, and at any given moment millions of them are rolling straight toward their destinations. On most days in most places so many waves are traveling the open sea that the surface appears to be a chaos of humps, peaks, and troughs. Waves ride on top of, crash into, bend and blend with each other. If the surface of the ocean could be instantly frozen, we could little by little unravel the groups of travelers. We would find many waves following one after another in evenly spaced lines called "wave trains," each traveling dead ahead toward its destiny.

If we could look at the wave trains separately, we would

see a uniform distance between the top of one crest and the top of the next. This is *wavelength*, just as it is for invisible radio, radar, and light waves. The vertical distance from the crest of a wave to the trough *behind* it is the wave *height*. The time lapse between one crest passing a given point and the next crest's arrival is the wave *period*. Disc jockeys on the West Coast often announce these three dimensions of coastal waves for the benefit of surfers. But even when the ocean is flat and the surfers leave their boards at home, wave motion has not stopped.

If anything is at all permanent and universal about the sea and the shoreline, it is the presence of waves. It is impossible for the world's seas not to make waves. Even a calm sea is only the placid surface of a wave too large for the eye to define. We can say this if for no other reason than that the tides themselves are waves. A wave is like a bulge traveling across the ocean's surface. The bulge is not water itself racing from place to place, but energy traveling like a ghost through the body of the sea. A wave may have birth, maturity, and death, but the energy it carries existed before the wave and lives on after its crashing death.

This distinction between water and wave energy becomes clearer if we think of a bather standing in the ocean just beyond the surf. He soon finds out that the easiest way to escape the push or breaking crest of a large incoming swell is to duck under the water. There, beneath the surface, he is rocked gently—downward and shoreward, then upward and seaward. If he is wearing a face mask and observes carefully, he will see that suspended sand, seaweed, and marine grasses growing in the sand are not rushed toward shore, but make the same orbital motion his own body felt. Of course these conditions are ideal. If tidal currents and conflicting sets of waves are present, the bather might find himself moved horizontally as well as in a circle.

The water in any wave, nevertheless, moves orbitally, and only the form of the wave or its energy content moves horizontally.

A wave is not a solid object like a boat or a fish. It is an intangible form or pattern of energy that shapes the tangible water as it passes, like a shiver in the skin. If the ship *L'Île de France* sails from Cherbourg, the exact same ship arrives in New York. If you could tag the molecules of water in a wave, however, you would find that any given set of molecules moved only a short distance, if at all, in the direction the wave moved. The water itself stays put. Anyone who has watched a piece of driftwood on open water has noticed that the wood does not ride on a wave but rises and falls as the wave form passes beneath.

Often ships can survive tremendous waves, simply rising and falling as the wave passes beneath. Thus, in intense storms the greatest damage done on a ship may be to make the passengers seasick. In 1933 a vast Pacific weather disturbance, lasting seven days, developed winds of sixty-eight knots and piled up waves scientifically recorded at over a hundred feet high. The freighter U.S.S. *Ramapo* got caught on its way home. The *Ramapo*'s officers were alert enough to keep the deadly waves astern as the ship wallowed forward, rising and falling on the waves, never opposing their gigantic force, and often riding down the steep front slopes like a surfboard. Had the ship turned broadside, it probably would have tilted and capsized on the slopes of the tall waves, or it might have folded into and been crushed by the breaking crest of a wave. By following the direction of the waves, the water's energy passed safely under the hull.

Despite their great variety of shapes and sizes, there are only three natural causes of waves: wind, earthquakes, and the gravitational pull of the sun and moon.

On the great waves of the tides ride all other waves and coastal storms. The tides themselves are created by the constant pull of the sun and the moon on the earth's waters, a pull which creates an upward bulge. Although the sun's pull is 150 times stronger than the moon's, the moon has the greatest influence because, being closer to the earth, it acts on much more specific areas of the sea. As the sun and the moon pull water toward them, the earth spins beneath its ocean blanket. This makes the bulge or tide travel across the earth—waves whose length is half the circumference of the earth. There is "no drop of water in the ocean," Rachel Carson said, "even in the deepest parts of the abyss, that does not know and respond to the mysterious forces that create the tide."

Tides are not the same the year around because the moon travels around the earth in an elliptical orbit. Once each month it makes its closest approach or its *perigee.* In this position the waters of earth feel its strongest pull and the wave of the tide responds by rising higher. Such tides are usually a minor swelling of the ocean's breast, a hardly noticeable witness to a quiet, ancient, and normal affair between earth and moon.

This ritual, which happens every 27.55 days, to be exact, occasionally coincides with a more familiar celestial event that happens twice every 29.53 days—the alignment of sun, earth, and moon. This position is called syzygy (siz-i-gee). When the earth is between the sun and moon, we on the dark night side of earth see the moon full face reflecting the sun behind us—a full moon. When the moon passes between us and the sun we see (or don't see) its dark side, the new moon. In either syzygy, or new or full moon, the combined influence of sun and moon causes the tides to "spring" upwards, a springing unrelated to the leaping growth of plants in March or April. Such spring tides occur twice each month, and like the swelling of the perigee, they attract little attention except from fish and fishermen.

But with mathematical certainty the 27.55-day cycle of perigee and the 29.53-day cycle of syzygy coincide exactly or call up the waters within a few hours of each other. At these times ten-foot tides become fourteen-foot tides. This is the *perigean spring tide*. And among these tides, there is still another special event. About once every year and a half, when perigee and syzygy are almost simultaneous, forces in the solar system temporarily warp the moon's orbit and bring its perigee closer to earth than normal. This unusual proximity is now called proxigee. During either the perigee-syzygy or proxigee-syzygy the moon can only maintain its distance from earth by speeding its flight. The tides rise faster and tidal currents run stronger.

Many storms which would have been relatively harmless by themselves occurred during spring tides. Such was the case of the great 1938 hurricane that lashed Long Island and New England and of the 1962 Ash Wednesday storm that tore up many resorts along the New Jersey coast.

Although most people do not think of the tides as waves, in certain places even the daily tides are shaped into breakers. This happens where the tidal advance is funneled suddenly into a narrow channel. One of the most famous of these places is in China's Tsientang River. Here, a sudden narrowing of the channel near Haining causes the tide to mount into a wall of water five to fifteen feet high. It comes roaring up the river as a single spilling wave and all small vessels must be beached. This tidal phenomenon is called a *bore*. The bore of the Amazon River is even more spectacular as it travels some two hundred miles upstream, becoming a roaring waterfall some twenty to twenty-five feet high. Since it travels at only twelve knots, the susceptible part of the river may at one moment contain as many as five crests of tidal waves.

Although tidal bores are uncommon and infrequent in the United States, several bays and harbors shape the tides in unusual ways. In the 150-mile-long basin of the Chesapeake Bay, for instance, there may be, in a given hour, the crests of two tides separated by a fifty-mile trough between them.

The great tidal shifting of water in the ocean expanses assumes little recognizable character until it touches the continental shelf and feels the influence of its tidal basins. Each basin is a section of the continental shelf whose particular shape, like a free-form bathtub, imparts a special character to the tide—a predictable height and a predictable schedule for the sloshing of the tide from one side of the basin to the other. The height, or "range," of the tide, as for any wave, is the vertical distance between crest and trough. In the tropical Pacific islands the tidal range is often less than a foot, perhaps because the islands are in the middle of the great Pacific basin and because there is almost no continental shelf for the tide wave to touch ground on. In the southeast basin from Cape Canaveral to Cape Hatteras, the tide sweeping into this great curve of the continent over a wide shelf rises from three feet in Florida and North Carolina to seven feet at the center of the basin in Georgia. At the land's edge its character is further formed by protrusions and indentations in the land, by the myriad inlets, bays, lagoons, and river valleys which it probes.

The continental shelf acts as a wedge driven under the tide wave, and like other waves, as the tide begins to rub bottom its crest rises. The Bay of Fundy in Nova Scotia lies before a large sloping continental shelf. It also lies at one end of a tidal basin called the Gulf of Maine, and the sides of the bay funnel the tide into a tight valley. Every day a 35-foot tide washes in so swiftly a man on foot cannot keep up running at full speed. When the tide recedes, boats anchored hundreds of yards offshore are left lying on their sides in the mud.

The cresting or subsiding wave of the tide also creates currents. Such currents reach ten knots in Seymour Narrows, Alaska, and four knots at San Francisco's Golden Gate. Complex tidal currents sweeping up Manhattan's East River and down Long Island Sound combine at Ward's Island off Astoria to create the treacherous Hell Gate, where many small-boat handlers have been swamped or stalled, unable to make way against the currents.

The largest and most powerful waves in the world are often sailed over by small boats without notice. Tidal waves, more properly called *tsunamis*, appear as gentle but rapidly moving swells when in the open ocean, but their energy reaches far down into masses of water, and when they approach land this energy rears up over the continental shelf in the form of huge breakers. This great energy is gathered from a rapid rise or fall in the seafloor itself an under water earthquake. The motion of the earth's crust displaces a mass of water, and the surface of the sea rises and falls with the new floor far below. From the quake's center a series of shock waves travels across the sea, yielding to neither wind nor other waves. Watching an incoming tsunami is like watching a normal breaker suddenly magnified to the scale of the legendary world of giants, a world where humans and human settlements are vulnerable as doll houses.

NORMAL TIDE RANGES

PACIFIC

Place	*Tidal Range (in feet)*
Anchorage, Alaska	26
Petersburg, Alaska	13½
Juneau, Alaska	14
Ketchikan, Alaska	13
Vancouver, B.C., Canada	10½
Victoria, B.C., Canada	6
Seattle (Madison Street), Wash.	7½
Port Townsend, Wash.	5
Vancouver, Wash.	1
Columbia River entrance, Oregon	5½
Tillamook Bay, Oregon	5½
Coos Bay entrance, Oregon	5
Crescent City, Calif.	5
Sacramento, Calif.	2
San Francisco, Calif.	4
Santa Barbara, Calif.	3½
Los Angeles outer harbor, Calif.	4
La Jolla (Scripps Pier), Calif.	3½
San Diego (Quarantine Station), Calif.	4
Honolulu, Hawaii	1

ATLANTIC

Place	*Tidal Range (in feet)*
St. Johns, New Brunswick	21
Eastport, Maine	18
Portland, Maine	9
Boston, Mass.	9½
Newport, R.I.	3½

New London, Conn.	2½
Woods Hole (Oceanographic Institute), Mass.	2
Bridgeport, Conn.	7
New York (The Battery), N.Y.	4½
Sandy Hook, N.J.	4½
Indian River Del. inlet	2½
Ocean City, Md.	3½
Virginia Beach, Va.	3½
Cape Hatteras, N.C.	3½
Cape Lookout, N.C.	3½
Cape Fear, N.C.	4½
Myrtle Beach, S.C.	5
Cape Romain, S.C.	5
Savannah River entrance, Ga.	7
Jekyll Island, Ga.	7
Mayport, Fla.	4½
Daytona Beach, Fla.	4
Cape Kennedy, Fla.	3½
Miami Harbor entrance, Fla.	2½
Key West, Fla.	1½
GULF OF MEXICO	
St. Petersburg, Fla.	2½
Pensacola, Fla.	1½
Mobile, Ala.	1½
Bay St. Louis, Miss.	1½
Ship Shoal Light, La.	1½
Galveston, Texas	1½
Atransas Pass, Texas	1½
San Juan, P.R.	1

The smallest, most varied, and most common waves, and those that have the greatest effect on the beaches, are born of the earth's fickle winds. There is no such thing as "airy nothingness." Air, of course, has substance and weight. As air blows across the ocean its molecules rub against the water, creating friction and drag. Anyone can recreate this process by blowing gently across a tray of water. The heavier water will not follow the wind any great distance, but it does begin to form crests, piling up and collapsing as the wind passes. The wind transfers some of its energy to the water. Once in motion, a wind wave moves across the sea, being reshaped, redirected, suppressed, or augmented by yet other winds and other waves. The final character of the wave we see approaching land has been shaped by three forces: the *velocity* of the wind; the distance, or *fetch*, over which the wind has blown; and the amount of *time* the wind has been pushing at the wave. The largest, most powerful waves occur when a fast wind has been blowing several days over open distances of a hundred miles or more. A wind of 15 knots blowing for two or three hours over a distance of ten miles will produce waves one or two feet high. But a 25-knot wind blowing for 15 to 20 hours over a distance of 150 to 200 miles will create waves 9 to 20 feet high. A large Pacific winter storm with 50-knot winds blowing several days with a fetch of 1,000 or 2,000 miles will pile up waves 100 feet high. The continental shelf and offshore bars of eroded beach sand tame these waves somewhat before they reach land by forcing them to break prematurely, but they still arrive with frightening force. Waves on the coast of Taft, Oregon, have thrown logs through the windows of the inn at Spanish Head.

A glance at the atlas will show us for any given coast what conditions are most likely to drive large waves ashore. Along the Texas coast southeasterlies streaming across the Gulf constitute the greatest threat. Along the New England coast it is the northeasterlies blowing across the long windy stretches of the north Atlantic, where storm waves created by sustained winter storms often reach heights of 45 to 50 feet. These waves, flinging pebbles and boulders and exploding against the cliffs, are slowly reshaping the continent's most durable shoreline. As on New England's rocky shore, the sharpest chisels of all may be the little air bubbles forced into cracks under tremendous pressure, which makes them explode with enough force to fracture granite.

To understand the many forms of waves and the way they work on the beaches, let's begin with an ideal wave. For practical purposes, all waves are variations on the ideal.

Let us assume that an ideal wave is born far out to sea—say, under the winds of a winter storm in the north Atlantic. For several days the wind has pushed and dragged at the surface of the ocean, its friction piling the water into waves. Our wave is born in one gust and nurtured by many more gusts. The wind blows strongly over a large fetch and the wave continues to grow, absorbing the wind's energy, which in turn came from the sun itself. Once out of the storm, the height of the wave subsides and the crest rounds. It becomes one of a number of swells moving beneath the peaks and froth of smaller, younger waves. Like all travel, the wave's journey has its dangers. Just as wind pushed it forward, a countervailing wind might wear it down, pushing the crest back toward the trough. Nor can all waves reach their journey's end through ice, rain, sleet, and hail. Sailors of the arctic ice pack may be in danger of being crushed, but even in gales few waves rise above the heavy ice. Even the lighter veil of snow or floating hail or the mild beating of rain gentles a rough sea. Oil, the modern villain of seashores, has at times blanketed a tossing sea and saved a boat.

Our ideal wave, being fiction, has safely passed all dangers and is now approaching the continent. If deepwater is usually the birthplace of waves, the surf zone, where waves expend their energy, is the graveyard. A wave dies in a few seconds, but in its dying it is spectacular and often dangerous. A good surfer, a prudent bather, and a wise property owner should be aware of what makes the apparently harmless energy of a swell become a spectacularly violent breaker. As a wave approaches shore it touches bottom and begins to break. Its orbiting water molecules have begun dragging at the bottom of their orbit, and the swell becomes a *shallow water wave.* This happens when water depth is about one half the length of the wave. In other words, waves with fifty feet between crests become shallow water waves at a depth of twenty-five feet.

As a wave drags on the bottom, its length shrinks, its approach speed slows, and the water molecules orbit more rapidly. The wave rises into a high ridge of water. Finally, when the ridge rises to a height about four fifths of the depth below it, there is no longer enough water in front of the wave to maintain a stable symmetrical shape. The molecules at the bottom of the orbit, dragging on the seafloor, cannot keep up with those on top. The crest of the wave falls forward and the wave begins to break.

A Pan Am Airlines ad shows a surfer riding beneath the curl of a huge plunging wave. This is the travel agent's dream, the *plunging breaker.* The plunger has a classic, sharp outline if unaffected by crosswinds or headwinds. Under calm weather conditions on a steep inner shelf the wave retains its stability and energy until the last moment. Then its smooth crest begins to fall like a waterfall over the trough in front of it. In the largest storm waves as the crest topples into the trough it traps a pocket of air and compresses it. Suddenly the compressed air breaks through the cap in a geyser that may leap like a fountain fifty feet in the air. In winter, along the coast of northern California and Oregon, plunging breakers thirty feet high pound the beach or outer sand bars with enough force to shake seismograph needles at stations several miles inland.

Most surfers, however, prefer a different wave and one that is constructive in its effect on the beaches—the spilling wave. A spilling wave is slowed down by a long, shallow inner shelf, bottom obstacles or local winds. Its crest, rather than racing before the wave's body and falling into the trough ahead, simply tumbles down the wave's front slope. Its breaking may last several minutes and span a distance of almost a mile. A steady supply of long, spilling waves has helped make Waikiki the world's luxury surfing capital. Waikiki's spilling waves are formed by a shallow offshore coral reef whose form is constant from year to year.

The higher the breaker, the more energy it will expend on the beach. The form of that energy depends on the shape of the wave. A steep high wave exploding on the beach creates a sudden backwash and sucks sand out to sea. A flat, tumbling spiller will spread itself over a greater area of beach and a greater portion will soak into the beach before backwash occurs. The sand that is suspended in this kind of wave is more likely to remain on the beach.

The final height and steepness of a wave cannot be determined by local weather conditions. A sunny, windless day at the beach can often witness long, high waves. On the West Coast, where long waves arrive from storms far out to sea, surfers at Santa Monica may be having a great day while sailboats lie becalmed just offshore, bobbing on the low swells passing beneath them.

A train of ideal waves traveling in an ideal sea would create a regular pulse of surf. Occasionally something like this happens and the surface of the sea is ribbed with long lines of waves. Stephen Spender, the British poet, has described a pattern of sunlit waves approaching land as strings in a harp.

> *There are some days the happy ocean lies*
> *Like an unfingered harp below the land.*
> *Afternoon gilds all the silent wires*
> *Into a burning music for the eyes.*

Seldom does the surface of the ocean show such simple patterns of waves following each other to shore. Usually it is shaped not by one wind or the *train of waves,* but by waves as diverse in origin as the crowds at Grand Central Station. In this confusion of interference waves beat each other down. Several waves may fall in step and reinforce each other. For a brief moment the energies of several wave trains unite and rise as one unpredictable giant wave. Ships sailing harmless seas can find themselves suddenly confronting waves fifty or a hundred feet high.

In February 1976, such a wave surprised the 16,000-ton transport U.S.S. *Shreveport* one hundred miles off the North Carolina coast. A stiff wind was blowing, but the officers on the flag bridge were not worried. Off the Cape Hatteras shoals at this time of year, however, the warm Gulf Stream, the Labrador Current, and cold northern air battle incessantly, creating a chaos of waves. Most boats, steering wide of the shoals, encounter no trouble, but just past midnight the officers on the bridge of the *Shreveport,* some fifty-five feet above the waterline, saw a foaming wave crest at eye level. As ship and wave closed with each other, the crest rose out of sight and a solid black wall of water smashed through the windows. No other giant waves followed, and the ship survived to take its casualties to the hospital. The *Shreveport* was more fortunate than the *Texaco Oklahoma.* In March 1971, the 632-foot tanker sailing off Cape Hatteras suddenly broke in half, losing thirty-one men, including all its officers. The most probable explanation for this and many other sudden open-ocean sinkings (including the disappearances in the "Bermuda Triangle") is giant waves. Just as supertankers have begun to ply the Gulf of Alaska, scientists using figures for the worst-known storms in the past twenty years have predicted insurmountable trouble. They conclude that the largest possible wave in the Gulf of Alaska could be 198 feet high. Call them what you will—rogue, freak, pyramidal—yachtsmen have the name which best applies to human perspective: "non-negotiable waves."

Granted, such waves are the extreme, reminders of what might happen. But consider the mass of water rising and falling in a few acres of ordinary ocean. Then the colossal energy expressed day after day is evident. Mankind, admirers of power and makers of machines, can barely comprehend the magnitude of the energy that expends itself on the shoreline. And few places in the United States have seen such sustained power as that which rages in the North Sea during winter. What more sobering experience than to have been at Wick Bay, Scotland, in 1872, when storm waves attacked a breakwater whose foundation was built of cottage-sized blocks weighing from 80 to 100 tons each, topped by three courses of large stones set in mortar, all topped by a monolithic cement cap weighing over 800 tons. Reinforcing rods three and a half inches thick pierced both the cap and the three courses of stone. The storm waves broke the concrete cap and the first course of blocks from the rest of the breakwater and pushed it inshore. Five years later, the rebuilt breakwater, now boasting a cap weighing 2,600 tons (or 5,200,000 pounds) was again broken by waves. An American engineer later estimated the force of the waves pounding against the breakwater at 6,340 pounds per square foot.

No wonder, then, that in 1971 when the city fathers of Sullivan's Island, South Carolina, constructed a jetty of old washing machines, plastic containers, bottles, and cast-off appliances, the waves very quickly turned their beach into a junkyard.

Measurements convey only a limited notion of the power of the sea. Sailors tell innumerable stories of great waves and stormy seas, but none compare with oceanographer Willard Bascomb's stories of the terrors of keeping a lighthouse on the open coast. Entire lighthouses from Maine to Alaska have been swept away by the waves. The Tillamook Rock Lighthouse in Oregon sits on a rock island several miles offshore, a little shoreline in its own right. The stone walls of the island rise some 90 feet out of the sea, and the light itself is 139 feet above the sea. During winter storms the whole rock island shivers, and boulders are torn from it and flung high in the air. One December storm threw a 135-pound rock clear over the light and through the roof of the keeper's house. Another time waves rolled a half-ton rock across the concrete platform supporting the lighthouse and destroyed an iron fence. Solid sheets of water 200 feet high crashed down on the roof of the lighthouse, and in 1912 a lighthouse keeper, investigating trouble with his foghorn 95 feet above the water, found it full of small rocks.

We have talked for hours with ocean watchers of all kinds—scientists, fishermen, yachtsmen, and the just plain curious and the tales of waves are unending. They are told with the unadorned, clear detail with which storytellers always honor people, animals, or events that refuse to be tamed or understood. Pacific storm waves like those at Tillamook Light moved Robert Frost to write these lines:

The shattered water made a misty din.
Great waves looked over others coming in,
And thought of doing something to the shore
That water never did to land before.

A casual observer might think that waves pushing sand shoreward and pulling it seaward in their backwash are the great carriers of sand. The back-and-forth motion of the waves stirs the sand, but over the long run the two motions cancel each other. Most bathers know that other forces are at work: the longshore or littoral currents which persistently carry them down the beach when they are trying to stay in sight of their own blanket. These currents are formed by waves approaching at an oblique angle. Part of their energy is deflected by the sloping sea bottom and veers off parallel to the shore, carrying with it both bathers and sand. Most of the long-distance moving of sand from one section of beach to another is the work of longshore currents.

The power of a wave and the angle at which its crest strikes the beach will determine how much sand can be moved. The ideal angle for the greatest transport of sand is about thirty degrees. The longshore currents created by the oblique waves become rivers of sand. By using special dyes and radioactive tracers, oceanographers have been able to follow the movements of individual grains of sand over many miles.

Table 2 shows the importance of the sand in these longshore currents, enough sand to build many beaches. Depending on the location, the sand may come from rivers, from beaches and cliffs, or from offshore deposits. It continues to move in the longshore current until the energy of the current is broken by inlets, submarine canyons, bars, and rock outcrops, headlands, or man-made structures. On the whole, the longshore currents move the beaches around but seldom produce a significant loss or gain in the overall area of beaches. It is one of the fairest transportation systems possible in a complex natural system—perhaps much fairer than the artificial transport of sand to selected beaches.

Table 2. Typical rates of sand flow due to longshore currents. These figures are from the U. S. Army Corps of Engineers, and most are derived by observing the volume of sand piled up behind engineering barriers on the beach.

Location	*Direction of Drift*	*Rates (in cubic yards per year)*
Sandy Hook, N.J.	N	436,000
Barnegat inlet, N.J.	S	250,000
Ocean City, Md.	S	150,000
Atlantic Beach, N.C.	E	29,500
Hillsboro inlet, Fla.	S	75,000
Vinellas County, Fla.	S	50,000
Perdido Pass, Ala.	W	200,000
Galveston, Texas	E	437,500
Santa Barbara, Cal.	E	280,000
Oxnard Plain shore, Cal.	S	270,000
Santa Monica, Cal.	S	1,000,000
Anaheim Bay, Cal.	E	150,000
Waikiki Beach, Hawaii	W	10,000

Along the wilder shores of the West Coast and wherever there are undeveloped national or state parks in the East, man seldom troubles himself about all the sand being moved by the longshore currents. A couple of feet a year taken from the north end of a beach and deposited on the south end threaten no one. In nature the sea does not seem to be eating its way inland at a rate that would bother any one generation.

On developed beaches our fear of the unknown often provokes us to meddle even when no danger exists. The sea is so powerful and unknown that if it moves even a few feet of beachfront, property owners want to reach out into the longshore currents and bring back their sand. Human beings have invented ingenious forms of breakwaters, groins, and jetties to weaken the currents and take from them their burden of beach material. No one can say with any legal authority who, if anyone, owns the sand in the longshore currents. Like any other resource that increases in value, it will certainly be claimed by more and more people.

The goal of man-made structures is to alter the flow of energy near shore and stop the beach from moving. For a short time this often works, and the beach assumes the shape we want. If it were all this simple, we would have engineered the ideal shape of our shoreline fifty years ago. Beach shape, however, is part of the dynamic equilibrium, and as it changes, it in turn changes the effect of winds, waves, and currents. After a while the structures built to meet one situation cannot cope with the character of energy acting on the beach.

5

The Beach Is Four Miles Wide

IN THE DAYS of the sailing ship many a sea captain, with his rigging in disarray and his ship rearing and plunging in heavy seas, sought shelter in the lee of a small island, only to find that, though the waves were decreasing in size, they seemed to follow the ship behind the island, bending as if to wrap the land in their crushing arms. Waves also bend shoreward where a headland juts out into the sea. The incoming lines of waves seem to bend upon the point of the headland and attack it from both sides. Old sailors used to say, "The points draw the waves."

While some mariners fear the coasts where waves bend around headlands and islands, the same bending helps make Waikiki's part of the Oahu Island shoreline a paradise for surfers. The great surfing beach is sheltered from the prevailing easterly trade winds and the headlong onslaught of their waves by the famous Diamond Head. As the waves approach the headland they bend around it toward Waikiki Beach. The waves, now approaching obliquely, begin to break at the eastern end of the beach where they are closer to shore. A surfer can ride sideways along a wave as its far end comes closer to shore and the breaking crest moves from east to west.

Powerful as waves are, something beneath the water guides their landward approach. Wave directions, heights, and periods all have different effects on any given shore, and each shore affects a given wave in its own manner. The waves bend around Diamond Head and other land forms because there is more to the shore than meets the eye. Beneath the waves actually lie two very different surfaces. The more or less stable, submerged shelf of the continent stretches from the continental slope far out at sea to the shoreline. On top of the shelf and moving forward with rising sea level is the *shoreface*—a broad, thin, band of restless sand and gravel, whose slope is much steeper than the almost flat shelf.

When we described ideal waves we spoke of the nearshore seafloor as a wedge driven under an incoming wave, causing it to slow down and increase in height. The shape of that wedge, however, is often very irregular. Although now submerged by rising sea level, eroded by constant wave action, and often loaded with sediment from the rivers, the continental shelf is a landscape that retains many of its old valleys, plains, hills, and mountains. The Santa Monica Mountains do not stop at the famous strand but extend many miles out to sea, making their last aerial appearance as the Channel Islands. Hills, mountains, and ridges in the seafloor mean that some parts of a wave line will touch bottom before the rest of the wave.

If one part of a wave touches bottom first, friction causes that part of the wave to slow down. Different velocities in different parts of the wave will cause a bending, or *refraction,* in the wave crest. Underwater shapes may weaken or intensify wave energy.

When a wave concentrates its energy on a headland, the

wave line touches bottom first in the vicinity of the headland, or rather on the underwater V-shaped prow of the headland. The point of contact with the prow acts as a fulcrum, wrapping the ribbon of wave around its sides. On both flanks of the headland the wave begins to pivot like a line of precision marchers, turning in toward the shoreline. California's Point Lobos and the other Pacific headlands are famous for the way in which they receive spectacular displays of the sea's energy. It is no accident that so many are the focus of state and national parks and recreation areas. Like lightning rods they make life safer in the adjacent coves or bays, where wave energy is low.

The opposite effect occurs where a deep submarine canyon cuts into the continental shelf near shore. Since the water over the canyon is deeper than the water over its sides, the wave's crest seems to stretch out thin over the canyon. The wave's energy disperses to both sides of the canyon, leaving a kind of "shadow" of relative calm at the beach at the head of the canyon. In this shadow sand drops out of the longshore currents, and shoals often build up.

Piers in California at Redondo Beach and Newport Beach are built near the heads of these canyons, where waves are rarely troublesome. For a short time shoaling was a problem at the Redondo Pier, and docking operations for fishing boats became difficult. One day in the early 1950s, fishermen noted the water deepening. After a few hours they had to let out an additional twenty feet of line to reach bottom. Shortly after that, the whole pier became noticeably shaky. Though everything remained quiet on the surface, a tremendous landslide of sand had flowed down the canyon; the annoying shoals caused by wave refraction had almost disappeared along with the pier and its fishermen.

Just how complicated refraction patterns may be is illustrated by the mysterious partial destruction of the breakwater at Long Beach, California, in 1930. A breakwater of boulders weighing up to twenty tons had withstood for years the Pacific waves and had protected the harbor. Then, between April 20 and 24, 1930, the harbor was singled out for unexplained violence. Huge waves attacked the breakwater and at its top dislodged stones weighing twenty tons. Three miles offshore, gambling ships reported good business as they rode calm seas. A few miles to the north at the San Pedro breakwater, there were no breakers at all. Along the southern beaches swimmers enjoyed the regular moderate surf. The big waves had come twenty to thirty seconds apart, a period typical of waves caused by great storms or earthquakes. Experts were mystified. Channel islands and the shape of the coast protected the harbor from any north or northwest waves. The waves had to have come from the southwest. For many years that was about all anyone knew: huge waves from an earthquake had come from the southwest. No one could explain how these could rear up at only this one point on the coast. Then, in 1947, a scientist studied the ocean bottom near Long Beach and found a peculiar hump, or underwater promontory, some twelve miles offshore and 250 fathoms deep. It acted like the headland of a shoreline, causing the enormous waves to bend and attack the breakwater like the inverted form of football's deadly flying wedge.

Underwater mapping may soon eliminate the surprises of events like the waves at Long Beach. Radar, sonar, and other electronic mapping equipment are rapidly making the topography of the continental shelf as well known as many land areas. Familiarity will make the shelf's effects on waves more predictable. The beach itself, however, can never be decisively mapped.

To most people, the beach seems to start at the low-tide

line and end at the line of trees or shrubs or cliffs. As if our interest stopped with vision at the water's edge, our concern and coastal planning efforts concentrate on the beach between the mainland and the low-tide line. This is the land most often listed in the assets of a development firm and outlined on property maps. This ribbon of sand, however, is only a small part of a broad, ever-moving expanse of coast that is the true beach. This true beach is actually the entire shoreface, a layer or wedge of sediment resting uneasily on the more permanent continental shelf, an insomniac on a firm mattress. In cross section it has the concave curve of a shallow saucer. It is constantly cleaned and scoured by the sweep of the waves. Perhaps because of its constant motion, the shoreface has only recently been recognized as a permanent part of the beach.

Even ideal equilibrium beaches come under the pronouncement issued by the ancient wise men who, when asked by their king to give him an eternal truth, gave him five words: *these too shall pass away.* All beaches lose sand. Sand settles out in the deep oceans, or blows inland from the beach, or moves laterally in the surf zone. Change in nature is constant. Yet the Taoists of ancient China are also right: *all change is illusion.* The dynamic system itself survives unchanged.

Sometimes the illusion is simply a matter of not being able to see from one season to the next. Some local governments are so committed to stabilization that they find themselves perplexed even by seasonal changes that are natural on beaches around the world. During most winters, residents of Virginia Beach, near Norfolk, Virginia, often see the broad beach in front of the dunes begin to shrink. On several occasions the city council has met in serious convocation to discuss the sudden change in the beach whose existence and intense use brings so much revenue and sustains so many businesses. Though these worries are reborn year after year, these seasonal changes are almost as reliable as the departure and return of migrating birds. The beach is merely changing its shape to survive the stronger winter waves.

Perhaps Virginia Beach residents would be heartened to consider the situation at Carmel, California, where a summer beach over two hundred feet wide retreats each fall, sometimes at a rate of six feet a day. By January or February there is virtually no beach at all. Farther south near La Jolla, broad sandy summer beaches sometimes in winter become little more than cobblestone strips hugging the edge of the rocky cliffs. Developers who understand this often work hard in summer months to establish their bulkheads and property lines far beyond the winter beach and well into the territory usually claimed by winter storms.

These changes seem dramatic and serious from the human point of view because they happen to the most cherished part of the beach—the *berm,* that sandy strip between dunes or cliffs and the mean high-water line. But the berm is only a small part of the whole picture, which continues both inland and seaward beneath the water. The berm is a flexible buffer, a kind of marine paunch which expends and contracts, storing and releasing the coast's diet of sand. Should the berm lack sufficient sand to meet the sea's challenges, behind it are the dunes or cliffs, vast warehouses of sand as well as obstacles to sap the strength of advancing waves.

In winter strong wave action chews into the beach and narrows the berm. Large waves often eat right into sea cliffs or carve a scarp in the first line of dunes. Some will wash over the first dunes into the back dunes. The berm vanishes altogether under strong attack, but only from sight.

The most visual and publicized seasonal changes in the

beach affect the narrow area washed by breaking waves, the "swash zone." This is the beach we usually walk on. The seasonal transformation that can change La Jolla's sandy swash zone to cobblestones is the result of stronger winter waves dragging sand out to sea. The beach continues to lose its fine sand until its materials are porous enough to absorb the destructive backwash of the waves. A fine sand beach absorbs very little water and thus the heavy backwash pulls sand with it, flattening the beach like a rake. The fine sand beaches of summer have a slope as gentle as one degree; medium sand, five degrees; and coarse sand, nine degrees, an angle approaching the slope of a steep street. A pebble beach may have a seventeen-degree slope. A cobble beach with a twenty-four-degree slope rivals many of the steepest streets in San Francisco.

A cobbled beach may be steep, but the increased wave energy that created it has actually spread the stolen finer material along the upper shoreface and flattened the whole system. On the East Coast, where there is often little difference in the size of beach sand available, the beaches from the first dunes to far out on the shoreface are always flatter in winter or immediately after storms. The steep, high-energy waves have only cut away beach sand to deposit it in deeper water or to a more protected place down stream.

The flattening of the shoreface slope is often the salvation of the beach. It creates a broad plain which constantly drags at the undersides of racing waves. Rather than breaking against the dunes or a narrow band of loose sand, the flatter shoreface scatters the waves' energy over a run that may be a few hundred yards long. Sometimes the beach counterattacks the waves with yet a second defense. Additional beach sand forms offshore bars or ridges with *runnels* or troughs between them. The presence of ridge and runnel systems makes incoming waves trip over themselves.

On Cape Cod, Provincetown and Truro are saved from the twenty-foot breakers of the North Atlantic by the annual formation of a large offshore bar made of sand eroded from the Cape. On the Northwest coast, where huge winter waves shake inland seismographs, hardly a single seaport or vacation colony would survive except for the offshore bars. From the cliffs in California and Oregon you can watch the great waves breaking a mile or two offshore and re-forming only half their original size. In warm weather the better swimmers and surfers mark the ridge's location as they gather there to catch the big waves.

Beach-property owners look with dismay at vanished strands and dunes sliced in half. After a bad winter or a hurricane, coastal residents and government protectors wax eloquent in their condemnation of nature's harshness. Plans are made to encase the remains of the beach in a seawall or to construct jetties and groins to grab back the lost sand. With few exceptions, the beach has not been lost. The vanished sand has simply been moved from dry storage to wet storage by steep waves and will be returned to dry storage by flatter waves.

In the Ash Wednesday storm of March 1962, Atlantic Coast residents thought they were seeing the ocean destroy the beaches from New England to Georgia. Constant winds and high tides sent the surf crashing into and over the dunes. On the Cape Hatteras National Seashore giant waves rolled in obliquely and carved away the dunes like a procession of swift bulldozers. Whole towns on the Jersey coast were wiped out, and even big resorts such as Atlantic City and Long Beach suffered millions of dollars' worth of damage. A month after the storm, many islands along the

Atlantic Coast looked like remnants of sand castles after a wave has washed over. Yet today most of the beaches are back where they were.

In mild spring and summer weather most waves do not have sharp peaks. They arrive as gentle swells, high but flat on top. These waves stir the sand in the bars and begin moving them closer and closer to shore until the bars are "welded" to the berm and form the summer beach on which we spread blankets and stake umbrellas. If you visit the beach early enough in the season, you can verify the welding process. Each day your swim from beach to bar will be shorter. Finally the bar will be visible at low tide, with a trough between it and the berm. When the trough disappears, the bar has been welded, the summer beach is complete.

So far we have been looking at the beach in cross section, as if we could slice through it from mainland to continental shelf. The changes in shape that are most worrisome to mankind, however, are the horizontal changes, the changes we would see if we were to hover over the beaches in a satellite watching the shoreline retreat, the marshes grow, the dunes travel, and the inlets wander. These horizontal changes are the changes recorded on most maps and aerial photographs. They are the changes which conflict with unchanging survey lines of private property.

Like a fighter under constant attack, the beach turns to face the waves. The glancing blows of waves that strike at an angle carry off more sand than do waves that come straight in. Thus, for survival, beaches try to face the waves. The waves, of course, change direction much faster than the beach can turn. The beach orients itself not to all waves but to those with the greatest power to move beach material.

What kind of wave moves the beach most depends on

the material. On the gravel and shingle beachs of California steep storm waves throw rocks into storm ridges which become the most durable part of the beach. The constant daily swells, while often large, are too flat and gentle in their washing to move these ridges. Thus, in this part of the country strong nearshore local storms create the waves which the beach must face.

On the sandy East Coast the flatter swells control the beach. Though born of storms, after hundreds or thousands of miles of travel, they approach land as long, flat waves, ideal for pushing sand ashore. These are the major beach-building waves. Local storms and steep waves merely erode the beaches. The direction from which the flat swells approach is the strongest influence on the orientation of the beach. There is one complication. Dominant waves generated in the Northeast do not necessarily mean that beaches will face that direction. Traveling over long stretches of the continental shelf, the waves are refracted by the shapes and contours of the seafloor and often approach from a direction that does not reflect their origin.

Now let us turn to the most popular beaches of all, along the barrier islands. On barrier islands, the beach, whose dynamic equilibrium is complex, becomes a small but vital part of a much greater equilibrium. The beach is the life source of islands which must actually roll over themselves to survive.

6

Barrier Islands: Let the Lighthouse Fall

In the mid-1960s, flying south from Norfolk, Virginia, the last city you saw was the crowded grid of Virginia Beach. Suddenly development disappeared and as far as you could see the foaming breakers rolled in on the thin tan-and-green ribbon of beach. A few years ago this barrier island was a wilderness beach. For over a hundred years people had lived there in small fishing villages and duck-hunting clubs, but the island of Currituck Banks seemed to have finally shrugged off the last of them. Like a fly on a horse, adaptable man had kept changing position, but finally fled.

The first challenge came in 1828, when New Currituck Inlet near the Virginia border silted in. Behind the island, Currituck Sound in two years became a fresh-water lake. Productive beds of clams and oysters sickened and died in the brackish water. Commercial fishing and shellfishing collapsed. Ocean-fishing fleets were locked out of their safe harbor. Residents began leaving. For a while, shooting vast flocks of migrating ducks and geese brought in money from northern markets. Rich Southerners and Yankees alike founded posh shooting clubs around the sound. As hunters killed hundreds of thousands of birds, the federal government outlawed market hunting. Private clubs withered when water milfoil began to choke out the eel grass and duckweed the game had thrived on.

Currituck's spectacular dunes also took their toll. In the 1950s 75-foot-high Penny's Hill began moving into the fishing village of Sea Gull. When Secretary of the Interior Stewart Udall flew over in 1964 and proposed a National Seashore for the area, the dune was pouring through the windows of the last house. Today parts of the village are emerging as ghosts of houses, and the dune itself, moving toward the marshes, has shrunk to a long ridge of sand.

In the great Ash Wednesday storm of 1962, surf rolled across part of Currituck and demolished the already dying village of Penny's Hill—just north of the dune that was at work burying Sea Gull. All along the island the sea washed over the frontal dunes. Currituck's rest from humanity has been brief. Local interests shouted down Udall's seashore while it was still an idea. A few years ago the developers began staking out lots on the very land scoured by the waves of the Ash Wednesday storm. Instead of wilderness, a visitor meets real estate salesmen, signs promising a beach paradise, and subdivisions flagged in thousands of small, expensive lots which often cost upward of $100,000 an acre.

Some of the new residents walking the beaches look curiously at the big chunks of fibrous black matting sticking out of the sand in the surf zone. Occasionally the stump of a tree protrudes, yet the only woods are up the beach, over the dunes and across the island by Currituck Sound's marshes. The black mat is peat and partially decayed plant material. This is the most certain sign that underneath the newly surveyed land the whole island is moving.

Currituck Banks has easily shrugged off human settlements not through a series of chance accidents, but because on barrier islands the dynamic equilibrium is not limited to the beach and the dunes. The whole island, from sound to ocean, moves with the beach, changing shape and position. Not only do all parts of the island move, but these smaller motions are part of a much larger one—the island rolling backward over itself, retreating in the face of rising sea level.

Geologists had suspected this giant rolling motion for many years, but a young botanist presented proof in the late 1960s. Dr. Paul Godfrey, of the University of Massachusetts, worked for the National Park Service on the Outer Banks during the summer. To look into the islands' evolution he began roving from lagoon to ocean, ramming steel tubes through the sand. He studied the cores the tubes withdrew. From layers of sand, bits of shell, pollen, and peat, he was able to draw accurate pictures of the island's anatomy, retell its history, and date the remnants of old beaches and marshes.

Godfrey has shown that between two and five hundred years ago the tree stumps and peat showing in today's surf were part of Currituck's back-side marshes and maritime forests, and that its beaches were a half mile or a mile seaward. Fifteen thousand years ago these islands were probably fifty miles seaward of their present position. A common view among geologists is that the barrier islands were born at the edge of the continental shelf, where it drops off toward the oceanic abyss. As the glaciers began to melt, the sea, four hundred feet below today's waves, popped over the edge of the slope onto the coastal plain. Rising rapidly, it began to chew its way into the coastal forests. Waves breaking on the old soil and flood plains pushed ahead of them masses of debris, mud, and sand. The wind winnowed out the sand and began to make dune ridges parallel to the shoreline. After a while the sea broke through the ridges and flooded the almost flat land behind them. The long, narrow pieces of surrounded dune ridge became the first barrier islands of the present era. Sea level was rising on the West Coast too, but barrier islands can only form on the nearly flat slope of a coastal plain. The few West Coast barriers are almost always sand spits, debris pulled off of weathering capes and headlands.

With water now behind and in front and tides washing in and out of the bays, a new set of forces controlled the sand ridges and the movement of the new islands. As sea level continued to rise, it drove the islands up the continental shelf. How fast a given island moves is controlled by how fast sea level rises and by the slope of the coastal plain over which the island moves. Along the Gulf Coast the coastal plain was so flat the islands must have moved one hundred feet a year at times. Naturally, the mainland shoreline behind the islands retreated too.

Some wide, durable islands, like Georgia's sea islands, enjoy a special, if temporary, state of limited grace. Cumberland, St. Simons, Ossabaw, and Jekyll Island, where President Carter hunts and swims, are naturally wider and rounder than their relatives in the Carolinas and Florida. The Georgia islands have migrated up to and are now part of an ancient barrier island system formed some thirty-five thousand years ago, the last barrier island before sea level began to retreat, marking the peak sea level of the last interglacial period. While some of these islands are more than a mile wide, the "modern" islands may be only a hundred yards across. In North Carolina near Cape Lookout the islands are just approaching the ancient ridge. Behind the Cape the elongate shape of the mainland towns of Beaufort, Morehead, and Harker's Island signal the location of the older barrier chain.

Since sea level affects both sides of an island, conserving

sand supply is the essence of survival. On the coastal plain there are no mainland bluffs or headlands from which the sea can mine new beach material. Rivers run slowly and drop their sand in the quiet estuaries. Only the migration of the barrier islands keeps them high enough on the coastal plain to stay above sea level. Since sea level is constantly moving, the islands cannot rest. The long, narrow islands in motion are, of course, more typical of barrier islands than the few which have united for a geological moment with the mainland or an island from another era. Standing free in the world's rising waters, these islands, no matter what coast they guard, have only three means of retreat—overwash, dune movement, and inlet formation and migration. We should be guilty of geological and biological stereotyping if we did not acknowledge regional and local differences. Each group of islands, like any threatened tribe, has developed its own common defense system. But within that system are local personalities of dramatic contrast.

Inlets, overwash, and dune movement all affect almost every barrier island to some degree, but the range of affect can be a life-and-death matter for mankind. The safest islands are those that have backed into older, thicker islands, as on the Georgia coast, though it is misleading to say that the "modern" part of the island is safe. The sand supply there is now very scanty, and the modern islands are often less than a hundred yards wide and extremely vulnerable to storms. The safe part is behind this narrow strip on the old Pleistocene island built some thirty-five thousand years ago when sand supplies were generous. These are technically "dead" islands. As sea level rises the islands will get narrower and narrower, until only the highest parts rise above water. Then, several hundred years from now, they will begin to move.

In the delta area of the Gulf of Mexico, barrier islands are often new, fragile, and fast-moving. A thousand years ago, as William the Conqueror was invading England, south of New Orleans Grand Isle was a sand spit detaching itself from a section of delta abandoned by the Mississippi River. When the river abandons a section of delta, erosion is rapid. Sea level appears to rise quickly as the delta's sediments compact and sink under their own weight. For the islands, sand supply becomes critical. The storm tides that top these low islands are harsh on development, but the frightening overwash actually allows the islands to survive.

The storm waves which roil the Gulf waters reach deep and bear great quantities of sand as they push over the islands. The new sand, layer by layer, spreads out in great sheets and fans. From this source the wind creates dunes, and new plant colonies take hold. Sand washed completely across an island into the marshes builds this area into new land. Pushed farther into the bay, the overwashed sand becomes a new marsh.

Overwash is so consistent and total on Grand Isle that development has been completely destroyed several times. As early as 1831 a fishing village disappeared under a six-foot storm surge and 150 people drowned. In 1893 the island lost its development again with 18 dead, a minor miracle when compared to the 1,100 people who died on neighboring Chenier Caminada. In 1915 storms in both August and September covered the island with more than eight feet of water, drowned all the livestock, and killed 275 people. People continued to rebuild, storm after storm. Jetties and bulkheads sprang up. Nevertheless, in 1965 Hurricane Betsy covered the island with a nine-foot surge, and all but three buildings in the thriving resort were demolished or heavily damaged. In 1976 jetties and seawalls had to be supplemented by sand pumping. But because the U. S. Army Corps of Engineers could not convince homeowners to move houses back (or simply remove them when no vacant land was convenient), the new beach was placed too far seaward. Its underwater slope was too steep, and after three months the new beach had washed away. Most of the new dunes behind the beach were too far seaward for grass to grow. Where the coastal Bermuda grass did take root, residents began to mow it, causing it to sicken and die. Along more than half a mile of eroding beach the Corps placed a longard, a sand-filled fabric tube, to act as a breakwater. The longard quickly sank to its midsection in the light sand and lost its effect. If these projects one day soften the effects of overwash, they will almost certainly lead to the island's death. Or else work will continue until Grand Isle changes from a resort to a fortress in the rising Gulf waters. On nearby Timbalier Island a rock wall, built to protect oil storage tanks, was simply left behind as the island retreated.

In the desert-like climate of the south Texas coast, wind, not overwash, is the most important provider for barrier islands. Padre Island is so long that the climate changes as you move south along its 130-mile beach, the air becoming hotter and drier, plants sparser, and their foliage tougher and thinner. The island is low and frequently overwashed even by spring tides and small storms, but the water is channeled by innumerable small dunes created by the wind. The flooding of major storms often does not create the general scale of flooding seen on Grand Isle because much of it flows through permanent storm surge passes, a phenomenon unique to the Texas islands. These passes are low, flat areas between dune fields, sometimes hundreds of yards wide. They function like inlets, but because of the small quantity of tidal flow in the lagoon, only the torrential rains and surges of major storms fill their channels.

Their dryness rather than their predictable but infrequent filling has led developers to site condominiums next to them. Here residents can enjoy a broad view out to the Gulf of Mexico uninterrupted by dunes. One developer on South Padre Island also built a wall across the storm pass. Luckily the wall fell in during a storm. If it had held, it would have backed up escaping flood waters from the lagoon into the condominiums.

When very high tides and storm waters fill the passes, Padre Island is divided into several islands. The phenomenon is so brief that we think of these separate islands as one system and the storm passes as part of the land itself. But what difference is there between a storm pass and an inlet which separates islands with different names? The easiest answer is to say the pass is often dry. This is a major difference to the driver of a jeep but not to the natural system. The storm pass is dry only because it is not called on by great volumes of water passing in and out of the lagoon with the twice-daily ebb and flow of the tides. Laguna Madre behind the island is so shallow and the climate so hot and dry that evaporation relieves the need for inlets by removing a significant amount of the water that drains from the mainland. Dense algae blooms frequently cover the water, and the small animals that are the basis of the marine food chain are stunted. Fishermen and tourist interests have promoted the trenching of artificial inlets, called fish passes, to bring fresh seawater into the near stagnant lagoons, but tidal flow is not enough to keep them open. They are no wetter than storm passes.

The great filler of the lagoon and builder of the island is the wind. Blowing across the dry dunes the wind constantly carries sand to the back side of the island, creating vast expanses of wind flats. This drifting sand makes Laguna Madre so shallow that on many days at low tide there is no water at all between island and mainland. The great flats hardly seem to be part of a sea island, but it is over this foundation that the more familiar dunes and beaches will retreat. And it is the broad expanse of the flats that protects the fragile desert island from back-side erosion.

On the Atlantic Coast true inlets may silt up and become dry land like the Texas storm passes, but their number and the volume of water they handle remain constant, a new inlet eventually replacing an old one. How high the daily wave of the tide rises along a barrier coast, and how much drainage enters a bay from the mainland, will prescribe how much water an inlet must handle or how many inlets must serve that section of coast. Along the Georgia coast in the center of the great arc of the shore that defines the southeast tidal basin, spring tides reach eleven feet. Here the inlets are frequent and the islands short. Along the Florida and North Carolina coasts, where the basin does not funnel such a great mass of the tide, the rise is only four feet. Inlets are less frequent and islands longer.

Before barrier island development became common, inlets seldom bothered anyone. Most development bordered the fixed inlets near the mouths of big rivers. Two hundred years of charts show inlets almost unchanged where the Altamaha of Georgia, the Cape Fear of North Carolina, and the Tombigbee of Alabama meet the sea. These inlets are the exceptions, the servants mainly of the great outpourings of fresh water in major rivers. Most inlets reflect the more changeable nature of the beaches and the sea. On North Carolina's Outer Banks, historian David Stick has located some twenty-six inlets now extinct but once alive long enough for their names to be inscribed on navigational charts. Inlets, however, are not as fickle as their erratic positions and comings and goings would indicate.

If we could measure the cross-section area of all the inlets draining a given sound, lagoon, or bay, we would find that despite the number and location of the inlets, they contained the same cross-sectional area—an unchanging ability to handle the continental drainage and the daily flow of tides. Inlets cannot be isolated from each other. They are interdependent. What does this mean for the island town or development that wants to close a bothersome inlet or for the fishermen and yachtsmen who want to reopen an inlet near their docks? If one inlet is closed, a new one must form somewhere else; if too many inlets are opened, sand will be inadequately flushed and one or more inlets will begin to choke. All this does not happen overnight. At first the system is out of balance. In a few months, a year, or more, however, dynamic forces move the beaches and restore hydraulic balance.

We introduced inlets as a third island-building force, but at this point the reader might say, with some truth, that we have only told how inlets move water. Moving water is part of the energy dimension of our equilibrium. Wherever water moves near the beach, sand also moves. The strong currents moving in and out of inlets cut across the longshore currents and break the river of sand moving along the coast. On the flood tide sand is pushed into the bay behind the beach to build new marshes and fill old channels. On the ebb tide it is carried seaward and settles in shoals or ebb tidal deltas.

The inlet is, in essence, a river with two mouths and, often, two deltas. On the ocean end its delta, or shoals, are low and scoured by waves. The flood delta formed in the lagoon not only fills part of the quiet water but attaches itself to the island. This is the first step backward in an island's retreat. The water above the delta is so shallow that marsh plants send down their roots. Big deltas sometimes have the appearance of marine prairies. Little by little the plants trap more sediment until land rises from the water and the island has moved backward and higher up the coastal plain, saving itself from rising sea level.

The flood deltas of inlets have made so much land that geologists believe as much as 40 per cent of North Carolina's Outer Banks rest on such areas. This does not mean that hundreds or thousands of inlets have opened and closed. All this land has been created by relatively few inlets. These inlets have created so much land because they wander or migrate. The sand pouring through an inlet often begins to build up on one side of the channel, forcing the main current of water to move and erode the other side. So long as the inlet does not meet great resistance or silt up, it will migrate. Like the pipe from a developer's dredge, the inlet fills the lagoon as it moves. The flood delta, however, does not rise above sea level until many generations of plants have come and gone, knocking from the wind its blown sand and trapping debris carried by the lagoon's currents.

Developers cannot wait for inlets to build new land, and in most cases they are not willing to give inlets room to migrate. Since some migration paths are several miles long, we can understand the impatience of the developer, but development near an inlet is risky at best.

As the first part of this book was being planned, we often illustrated our ideas to friends and students by taking them to Bogue inlet at the end of the Emerald Isle resort development on North Carolina's Bogue Bank. Bogue Bank parallels the sudden and short westward turn the coast makes after the point of Cape Lookout. Thanks to a good sand supply and a slow rise in sea level, the island was expanding until about a hundred years ago. Most of the island is still high and wide. Its relative security has attracted developers who until recently admitted no danger anywhere. On the west end the main road of the Emerald Isle, subdivision stopped in a cul-de-sac almost at the edge of Bogue inlet. Prospective buyers were lured by the water view and the frequent passage of colorful shrimp boats through the inlet. One day in 1972 someone noticed that the inlet had moved closer to the new cul-de-sac. Buyers took little notice and began building houses on nearby lots. One lot owner constructed a small boat dock near the inlet, only to have it collapse into the moving channel the day after completion.

By August 1972, the inlet had cut into the cul-de-sac itself, as if intent on following it into the subdivision. In late August we took a Duke University geology class to look at the ruined road and study the inlet, which was then chewing its way into the island at a rate of six feet a day. A real estate salesman passed by with some clients, explaining to them the great investment potential of the remaining lots. The background noise for his talk was the quiet *plop, plop, plop* of the inlet's banks slipping into the high-tide currents.

The Bogue Bank inlet was not on the warpath, as local and state papers sometimes described it. The inlet was simply doing what many inlets on barrier islands do—migrating. Ironically, the developer may have helped the inlet migrate faster, since some evidence suggested that the new road had been built with large quantities of sand mined from local dunes. The inlet may have followed the path of least resistance by eating its way into the weakest part of the island. In any event, old maps of the area show that the same inlet has from time to time migrated up and down the beach. Emerald Isle is now so highly developed that wherever the inlet moves it will be condemned as a "natural disaster." But the subdivision is also incorporated as a town, and everyone is being taxed to pay for stabilization efforts that the U. S. Army Corps of Engineers' experts feel are futile. A prominent bank executive solved his problem with the inlet by simply giving away a whole block of development land, most of it already underwater, to a private school and deducting the gift from his income taxes. Conservationists who knew the inlet and the development saw the gift recorded in the county's deeds. They put two and two together and called the IRS. Their eventual reward enriched the Conservation Foundation of North Carolina by over five thousand dollars.

Among the documents they presented to the IRS were old maps showing that Bogue inlet, like most inlets, wanders and that a reasonable person would have known the donated "land" was worthless. While the wandering of inlets robs developers of whole subdivisions and endangers many homes, the movement is part of their function in bringing new sand to the islands.

Most inlets sooner or later are so successful in their constructive work that they destroy themselves. As the tidal deltas grow larger, flushing becomes more difficult, even under the pressure of storm tides. This is the time when, having finished their constructive island building, inlets must close.

Moriches Inlet west of Fire Island National Seashore in New York showed its age by closing naturally in 1950. It opened again under storm pressure in 1953 but has been silting in ever since, although the Corps of Engineers maintains jetties to stabilize the channel. In 1978 the corps proposed lengthening the jetties and dredging. The inlet, however, has built such large ebb and flood tidal deltas that it would be easier to create and maintain a new inlet at one of the island's narrower points nearby.

When such aging inlets are left alone the water which

must move in and out of the lagoons eventually cuts another inlet, leaving the old channel to fill itself and become part of the island. The filling of an inlet may take five years, fifty years, or several hundred years, but the aging process is hurried if people try to wall up the inlet to keep it from migrating. Channel dredging maintains a safe passage for boats but seldom preserves the inlet's flushing ability. In our shortsighted concern over migration and boat channels we often cannot see far enough up the beach to know that we are encouraging a new inlet to open.

On Florida's Treasure Island off of St. Petersburg, the concern was not so much navigation through Blind Pass as the fact that in 1931, when real estate speculation was just beginning, Blind Pass was closing up and a new inlet had severed a hopeful developer's holding from the rest of Treasure Island. The developer bought out an automobile junkyard and filled the new inlet with the old hulks, then hired a dredge to pump in sand so he could build cottages on the new land. The project was a success for the developer, and Blind Pass is still used today. Flushing is so difficult that numerous jetties and dredging operations have failed to satisfy boat owners. The worst result is yet to come. Blind Pass could be one of the silted-up inlets mentioned in John MacDonald's best-selling novel *Condominium*, and heavily populated Treasure Island could be the island cut in half when storm surge carves a new inlet through a narrow, denuded section of island, toppling several high-rise condominiums loaded with senior citizens. Dumping cars in nature's replacement to Blind Pass will one day make reality's disaster greater than the novelist's invention.

The marine civil engineer in the novel, who predicts the opening of the new inlet and the falling of the condominiums, makes such prophecy sound simple. In fact, there is no great mystery to such prediction, though often we'd rather not think about it. The warning about building near an existing inlet could also be expanded to a warning against building near the low, narrow part of a barrier island where new inlets are likely to form. This warning is particularly important on a developed island where dead-end canals and marinas essentially bring the ocean and bay closer to joining. Artificially closed inlets are also candidates for reopening. When not paved or overgrown, their presence can sometimes be detected in the dark color of shell fragments in the dredged sand. At Wrightsville Beach, North Carolina, the high-rise Holiday Inn sits on top of an old inlet, and the sand to fill the inlet was dredged from the tidal delta immediately behind the motel. Where once a delta made this part of the island thick and safe, it is now narrowed by a deep hole in the bay where roots of marsh grasses cannot anchor. During a hurricane the enlarged mass of water in the bay may rise and cross to the ocean. Local watermen and knowledgeable geologists and planners refer to the motel as the Holiday Innlet.

Inlets have earned respect for their role in building flood deltas, but their constructive work does not end there. While acting as escape valves for storm waters trapped in a bay, they also help preserve the open beaches. With the storm water that rushes seaward through an inlet come massive amounts of sand from the bay. This is not sand that is going to be totally lost out to sea. Just as in normal weather the inlet carries sand to build the back of the island, so during storms an inlet brings out material to rebuild eroded beaches. How can this be if inlets only build islands over months and years? Like any other part of the island that adapts to high energy, an inlet can transform itself quickly. During a storm its channel widens and deepens almost catastrophically. An inlet knee-deep in gentle weather might be twenty feet deep during a storm. The emergency relief of the inlet's growth not only saves the island by releasing flood waters from behind it, but is the first step in preparing for natural restoration with sand flushed out of the bay.

Perhaps the most important lesson of the shifting inlets is to remind us that no one barrier island can survive by itself or even maintain its single identity for very long. The islands are only defined by inlets, and inlets by nature make that definition temporary. The long chains of barrier islands—the New Jersey chain, the Maryland-Virginia chain, the North Carolina chain, the South Carolina-Georgia chain, and the Florida chain—are single geological units like the pieces of a necklace. They share a common supply of sand and respond to the same body of water.

Most of the sand that makes barrier islands at one time or another crosses the beach. If the prevailing wind blows onshore and the island's climate is favorable, beach sand is blown and washed into the piles we call dunes. Many islands have neither favorable winds nor climate. Until coastal land became very scarce and expensive, man favored islands with extensive dunes, even if they were quickly bulldozed. These are the highest islands, and in the initial modest stages of development dunes are kept as protection. This modicum of respect may be because we can easily see how dunes blunt the force of storm waves. In our wiser moments we have even hidden our hotels and houses behind the dunes rather than push them out of our way in order to get closer to the sights and sounds of the ocean. The existence of a dune system relies on two big ifs—favorable prevailing wind and good sand supplies.

The survival wisdom that makes dune systems the heroes of the natural process embodies a mistaken idea of the dunes' importance to the islands. This "wall view" has led us to spend many millions of dollars creating artificial dune walls that look natural to the untrained eye but which are in fact destroying the beaches and setting the stage for major disasters. The almost unanimous praise of dunes as a natural wall that stands against the sea rests on a limited and very human bias, one that becomes clear if we ask what maintains those duneless islands that are frequently overwashed.

In the center of the North Carolina coast the elbow of Cape Lookout marks a turn in the barrier island chain, the location of two adjacent but drastically different islands. Core Banks runs north and south. At its southern end the coast turns at the Cape and Shackleford Bank stretches westward. Winds from the south strike Shackleford head-on and are constantly blowing sand from its beaches into a high and broad dune system that shelters a large maritime forest. As they travel north the same winds blow lengthwise along the beaches of Core Banks. What few small sand hummocks exist on this stark and narrow island get scanty supplies from the beach. Both islands share the same sand supply, but the sand that reaches Core Banks beaches is seldom blown landward above the high-tide line. Core Banks, like many islands, has found another way to survive. It thrives on overwash.

Even a casual look at a natural dune system will prove that nature does not build solid walls against the sea. Numerous gaps allow wind to blow through and storm tides to pass. In the natural system dunes are not so much defenders of the island as warehousers and distributors of sand. They are born of the dynamic equilibrium and move with it.

However majestic in their breadth and height, dunes have humble beginnings. A clump of seaweed, a log, or an abandoned bottomless rowboat traps a little hummock of sand. If grass or a shrub begins to grow, a dune is born. Sand blowing in the wind strikes the leaves of the plant and falls into the new dune. The plants in turn send roots out to hold the new sand and rise just a little higher. As a dune system develops, the sand it accumulates is shared by the entire island. Sand blown off the frontal dunes drops out in their lee. The middle and back of the island are built up and secondary dunes form. Sand spilling out of the toes of the dunes builds up the interdune areas. As some dunes migrate they leave behind sand which is colonized by new communities of plants. The result is usually a high, thick, and relatively safe island. But it is an island that still moves.

As sea level rises the dunes back up. The movement is complex and hidden in many surface changes. It shocks beach residents to see storm waves carve a scarp in the front of protective dunes, but this is a normal part of the system. During the storm the sand yielded by the dunes helps flatten the beach and forces the waves to break farther from shore and dissipate their energy over a longer run. In calmer, drier weather the waves return the sand to the beach and the wind lifts it back into the dunes. The system is restored, although the dunes may, in fact, have retreated a few inches.

To property owners who have built immediately behind the front dunes, the sand blown and washed back often fills their carports and storage areas. After many winter storms and hurricanes they must dig themselves out. At the same time county highway machines are trucking off sand that has covered the streets. Along with the sand goes the island's attempt to strengthen its midsection against the time when the frontal dunes fail.

The biggest, fastest-moving dunes of all are sometimes

far from the ocean's waves. Earlier we mentioned one such dune snuffing out a little village. Farther south the famous 150-foot-high Jockey's Ridge has been slowly wandering down the back side of Hatteras Island, its steep sides spilling over whole forests and marshes. In its wake it has left the island higher and safer. A few years ago developers were beginning to bulldoze the sand mountain to make way for a housing development. A furious campaign by local residents, conservationists, and hang-glider enthusiasts convinced the state to buy the dune.

The existence of great dunes like Jockey's Ridge take us back to the beach, to the strand, the berm, the narrow strip of bare sand whose image most quickly comes to mind when we use the word *beach*. On mainland coasts where the beach often stops suddenly at a cliff, a bluff, a forest, or a highway, it is easy to think of a beach as a certain amount of sand. It may come and go, shrink and expand, but its fate seems to have little effect on anything else.

Barrier islands and barrier spits, with water behind and before, could not exist without the beach. Marshes, giant dunes, forests—they are made and constantly remade out of the beach sand. The well-being of the beach is the very existence of the island. The islands have existed for fifteen thousand years, but not in the same place or of the same substance. Only the form and process of the islands has existed that long. The trees, dunes, marshes, and the exposed sand itself may be only a few years old, at least in their present appearance and position. The peat mats and stumps which appear in the ocean surf are sometimes less than two hundred years old, telling us that here an entire island has rolled back over its old forests and marshes since the Declaration of Independence. Most of the houses, hotels, condominiums, and stores from Atlantic City to the

Mexican border are built on sand that blew and washed in from the beach, much of it only five or ten years ago.

What places us in most danger is not the gradual rolling over of the barrier islands, but our insistence on occupying and stabilizing, buying and selling every valuable square foot of real estate. That short-term preoccupation leads us to spend most of our energy fighting and quelling the very forces which ensure an island's survival. We value only the material part of the four elements of the dynamic equilibrium. And we value only the material that stays on our property. We occupy the barrier islands like people who have built cities on the backs of giant sea turtles. Like all reptiles, turtles must breathe, but we so enjoy our perch we will not tolerate the beast moving even to keep a footing in water shallow enough for it to raise its head.

Some 280 barriers protect the U.S. coastline. Seventy are already strait-jacketed with heavy development. A hundred more are being developed. All coastal states have coastal-management plans, but not one recognizes the indivisible wholeness of barrier islands or plans to designate barriers per se as areas of special concern.

Perhaps as a nation we are too traditionally optimistic, confident of avoiding or conquering all disasters. We choose to stand, express ourselves freely in our development, and fight the natural forces with our engineering genius. The barrier islands, as their name implies, have barred the sea from the mainland. We have become so confident in our technology that engineers now propose their own barriers to bar the ocean from the islands.

7

The Human Nature of Natural Disaster

THERE ARE NO catastrophes or disasters in nature. The powerful events that seem so unusual to us are disasters only because our lives are short and our bodies fragile. We define disasters by contrasting them to familiar, frequent, and harmless events—the fall of night, a rain shower, birds singing. Unfortunately, most of us take little note of ordinary phenomena. And since nature per se is no longer the home of the divine spirit, we disassociate its daily forms from the spectacular or the extreme. These disturbances come upon us, not as a part of the continuous whole or events integral to the drama of nature, but as mutations, intrusions, freaks, or "acts of God," a God who stands apart and turns nature ugly when his mysterious purpose moves him.

Great storms, floods, landslides, and quakes are, in the grand scale of time, normal events recurring again and again, more or less regularly. Lightning, if we could watch long enough, strikes the same place many times. While thousands of humans perish in a single stormy night, the redwoods survive two thousand years, and some mountains have witnessed millions of years of wind, ice, and sun. A fair consideration of natural disasters must begin by admitting that whether an event is a disaster is a matter of the size and longevity of the ox that's getting gored.

Whatever governs the order of nature's events, the greatest events themselves can be observed and explained without setting them apart as acts of God. "Disasters" and "catastrophes," whether predicted by prophets or decried by the average citizen and sensationalized by the media, are actually sudden and local concentrations or transformations of natural energy—the same energies that are always at work in nature, the same energies that have shaped land and sea for several billion years.

Every day parts of the southern California cliffs yield to gravity, even if only a few pebbles fall. But the falling, like gravity, is always there, and how much falls depends on how the kaleidoscope of natural events arranges itself. In wet weather and a high surf the falling of a few pebbles becomes magnified into a landslide. So too, the gentle winds that circle the high-pressure systems that normally cross the continent are harmless cousins of stronger cyclonic winds—northers, squalls, northeasters, spring storms. And by whatever name hurricanes go by—typhoons in the Pacific, willy-willy in Australia, tropical cyclone in the Indian Ocean, chubasco in California—to a meteorologist they are all migrating cyclones.

On an undeveloped coast one might react to a landslide as some parents react to a daughter's marriage: we are not losing our land but gaining a beach. An island's dunes overwashed by a hurricane give up their sand to form new marshes. "Disasters" are part of the dynamic equilibrium of the beaches.

This sense of proportion is hard to maintain if your

house is teetering on the edge of a cliff, has been washed off the top of a dune, or smashed by storm waves. Natural events certainly create personal, social, and even political disaster. Americans, being friendly, assertive, and confident people, are insulted by nature's attacks. Our instinct is to stand up and be counted, to brave it out and strike out. After Hurricane Camille in Mississippi, American flags broke out everywhere, many of them flying over bare foundations. Too often, of course, those who dared hurricanes and landslides have been counted lying down dead.

Storms, landslides, and tsunamis, despite their destruction, offer a valuable frame for human activity on the beaches. First, they provide a measure of the ultimate forces we will encounter. Second, because they can move the beaches in a matter of hours, they often tell us just how transient that landscape is. Like a painting in a good frame, human activity in the context of natural disasters appears with new clarity.

Landslides

We think of water exerting its shaping force on a land that passively endures. This order of nature occasionally reverses itself when a large piece of the land falls with spectacular, if local, results. A falling land mass not only reshapes the beach on which it falls, but it places in the waves and currents a valuable supply of material for beaches near and far. All along the Pacific shore landslides have made and will continue to make important contributions to shoreline evolution.

As beneficial as they are, riding a landslide down to the beach or into the ocean has never been suggested as sport, though Californians seem to have no fear of the ride.

Houses abound on sea cliffs, some cantilevered out over the cliff face. The closer to the beach you are, the more prestige you have. And many Californians are lulled by a beneficent climate into a naïve sense of kinship with nature. Being close to the beach in this case, however, means eventually becoming part of the beach—something no house survives satisfactorily.

The worst conceivable situation for landslides would be a mass of seaward-dipping sedimentary rocks perched on top of a layer of weak, water-permeable shale in a wet climate. On the other hand, a granite mass with few fractures or joints, on a desert coast, would be comparatively safe. The safest place of all, however, is several hundred yards or more inland from the shore.

The response of many Pacific Coast residents to the predictability of slides is to ignore the threat before it happens and to suffer afterward (if they are alive). In the Palos Verdes to Santa Monica area there are many demonstrations of man's willingness to ride landslides to the sea. On the south coast of the Palos Verdes Hills near Portuguese Bend, the cliffs are weathered volcanic tuffs (rocks of old volcanic ash) interlaid with muddy sediments. The tuff beds slope seaward, an ideal condition for slides. The clays swell and become soft when wet. One slide in this area extended along the coast a mile in length and caused abandonment of an entire residential subdivision of expensive homes.

Farther north at Pacific Palisades people have perched their houses on top of and below unstable sedimentary cliffs. For years the coastal highway along the beach has been troubled by collapsing shoulders and falling debris, and taxpayers have kept it in repair. A state highway engineer was not so lucky one day in 1958. As he was about to open a repaired stretch of the coastal highway, 600,000 cubic yards of mud and rock fell. He was buried on his road as the slide passed over it and pushed out into the water.

At Santa Barbara near the University of California at Goleta, expensive homes are built right up to the very edge of the vertical cliffs formed of weak shale and siltstone. Some are cantilevered over the beach or stop flush at the edge of the cliff. Sections of cliff in this area have eroded one to two feet a year, and an earthquake of 5.3 on the Richter scale in 1968 and another in 1978 hit the area, casting pieces of cliff to the beach below and warning that the sunny day of the development was drawing nearer to a close.

The companionship of earthquakes and landslides may be why Californians and Alaskans disregard the warning rubble on the beach. Severe earthquakes are still unpredictable and arrive years apart. Many sections of the West Coast have not had large landslides since the 1906 San Francisco quake. Perhaps if you're going to risk the quake, you might as well risk landslides too, although that is asking more than double trouble. Falling down a cliff or being buried by tons of rock and earth guarantees almost a 100 per cent fatality to house and owner.

In the March, 1964, Alaskan quake 60 per cent of the damage in Anchorage was done by five large landslides. Along the coast great blocks of earth that had been stable for years slumped seaward as loose gravel and clay underneath them gave way. Schools, government buildings, and whole subdivisions became a tangle of broken walls, severed roads, and deep ravines. The destruction in Anchorage is a small picture of what will undoubtedly happen any day along the heavily populated California coast where thousands of houses, apartments, and public buildings sit on the bluffs waiting. From San Diego to Mendocino former slumps and landslides are evident, though the edges are blurred by vegetation and erosion. Where the cliffs fell before, they will almost certainly fall again, because falling is built into their geology.

Landslides on most U.S. shorelines are impressive and dangerous, but not nearly so far-reaching in effect as the landslide that transfers its energy to water. When this happens a surge or local tsunami is formed, spreading the energy of the slide far from the actual site of the event.

The cold, quiet fiords of southern Alaska are the scene of the most spectacular landslide surges in the world. The mouth of Lituya Bay north of Juneau is narrowed to 1,000 feet by a hilly spit. Behind the spit the deep-blue water stretches eight miles inland, rimmed by the close embrace of high mountain walls. In the mountains at the head of the bay is an active geologic fault.

On the evening of July 9, 1958, three fishing boats were anchored in the harbor. At 10:16 P.M. the sun was still shining in the Arctic night. At that time the boatsmen felt the tremors of an earthquake, nothing uncommon in the seismically active subcontinent of Alaska. But two and a half minutes after that, they heard from the head of the bay a roar like something out of the Apocalypse. They were too far away to know that 40 million cubic yards of rock (about 90 million tons) had fallen into the bay from a height of 3,000 feet. And if they had known, they would have had little time to care, because advancing on them three minutes later was a mountainous water wave 100 feet high.

One boat snapped its anchor line and rode the edge of the wave as it washed far up the south side of the bay. It would have crashed on the rocky hills if the backwash hadn't returned it still afloat to the center of the bay. There its occupants rode out a series of later waves some twenty feet high. The first wave also picked up a man and his wife on their boat, and they found themselves surfing across the spit at the harbor mouth. Some eighty feet below them the tops of trees growing on the spit passed by swiftly. The boat came to rest in the ocean, and the man and his wife escaped the swamped and sinking craft in a dinghy. A third boat carrying two passengers was never found.

Observers flying over the site the next day noted that the wave had cleared away all trees up to a maximum height of 1,740 feet above the bay. Above that the bark was frequently stripped from the trees. Where the wave hit the trees directly, it cut a trimline as neat and sharp as any glacier would have done. Where the soil was deep, trees were washed away, roots and all, while even large trees anchored in rock were snapped like matchsticks.

Yet these were not the first giant waves to hit Lituya Bay. In October, 1936, three waves of 50 to 100 feet had swept down the bay, and scientists have found evidence of other great waves (in 1853, 1874, and 1899), all ranging in height from 80 to 400 feet. When the bay was discovered in 1786 by a French expedition, a small Indian village thrived on its shores. The village was short-lived, and the bay has not been inhabited since. It is known chiefly for having the highest waves ever witnessed by mankind.

Lituya Bay remains as a classic warning to settlers who build in the fiords of earthquake-prone Alaska. Its warnings, however, have not been strong enough to overcome the hopes of politics and economics. During the 1964 quakes towns built in fiords or at the mouths of fiords were devastated by slide-caused surges. The town of Valdez, where oil from the Trans-Alaska pipeline now pours into supertankers, sat originally at the edge of an unstable delta deposited by rivers emptying into a deep fiord. When mudslides occurred at the delta's edge two hours after the quake, 20-foot-high waves traveled the entire twelve-mile length of port Valdez fiord and back again. After studying the devastation, government officials decided to rebuild elsewhere. The thriving supertanker port of Valdez has been moved four miles west. It is now safe from slides, but both tankers and tanks are much more vulnerable to tsunamis.

The fate of large oil-storage facilities and of tankers during a quake or tsunami is something that would correctly be called a *natural* disaster, but only because in the resulting oil spills and fires nature as well as humanity suffers enormous damage.

The same 1964 quake provides a preview of such a catastrophe. It occurred only a few miles west of Valdez, at Seward at the head of Resurrection Bay. This was the scene of another delta mudslide from the quake. The resulting local tsunamis carried boats in the harbor high over the breakwater and out into the bay. Many people were fortunate to escape to higher ground on foot and by car immediately after the end of the quake. Twenty-five minutes later, those who were left in town saw waves 30 to 40 feet high spanning the width of Resurrection Bay, sometimes covered with burning oil. The oil had spilled from tank cars at a Standard Oil dock. One flaming wave caught a slow-moving train of oil tank cars, igniting one after another until the entire train was in flames. The quake left thirteen people dead in Seward. Eighty homes were completely destroyed, many others severely damaged.

There were no nearby alternate site on which to rebuild Seward, and today, after spending $22 million on reconstruction, it stands as before, an important transportation center at the head of Resurrection Bay, a sitting duck for future disaster.

Perhaps those who wrote the Book of Revelation conceived of that Apocalypse by watching human willingness to set and reset the stage for innumerable rehearsals.

Tsunamis

Travelers in the open stretches of the Pacific Ocean have ridden out the most powerful waves ever known without putting down their cocktail glasses. They may even have sailed peacefully over the colossal earth movements that generated those waves. Although authorities disagree on how tsunamis are formed, most feel that the waves are born of some kind of underwater earth movement, sometimes a landslide, sometimes an earthquake. Tsunamis are long-period (time between passage of one wave crest and another's arrival) waves, with lengths sometimes exceeding one hundred miles. Sometimes crossing the entire Pacific Ocean, they travel at speeds up to five hundred miles per hour in the open sea where, fortunately, their height is seldom more than a foot or two. Their depth, or the reach of their orbiting particles, however, is thousands of feet, and they touch bottom even far at sea. When approaching shore, tsunamis react like other waves. As they encounter the marginal slopes of a large island or a continent, their waters begin to pile up, and the waves that crash over the land may reach a height of one hundred feet.

Although there are hundreds of underwater earthquakes each year, tsunamis are rare. The ideal condition for a large tsunami is a strong earth movement occurring within 150 miles of the sea. While tsunamis contain more than a hundred times the force of the Lituya Bay landslide surge, they nevertheless have only one tenth or one hundredth the energy of the seismic waves that give them birth and sometimes send them from the Aleutian Islands all the way to Australia.

In deepwater the energy of a tsunami expresses itself in speeds of hundreds of miles per hour, and its force remains undiminished across vast distances. Laboratory models have suggested that tsunamis arriving on a gentle sloping shelf will not rise as high as those arriving on steep slopes, because the friction of the gentle slope dissipates some of the wave's energy as it moves landward. A shoreline at an oblique angle to the oncoming wave also benefits from increased offshore friction and does not receive as much energy as a perpendicular shoreline. Tsunamis, however, can be magnified by other special conditions as well, such as harbors and bays. A triangular bay can squeeze or funnel a tsunami into a wall of water. A wave that might harmlessly break on the open beach could devastate a town at the head of a "snug" harbor.

Within minutes tsunamis shape and reshape entire beaches and shorelines, but their most permanent damage is to human beings and their structures. Since the Greeks first recorded tsunamis in 1400 B.C. in the Mediterranean, tsunamis have occurred with increasing frequency, though this may only reflect the increasing numbers of people who live along vulnerable shorelines and report the waves. All coastal civilizations record with awe the visitations of great waves, but nowhere are these waves more fearsomely and elegantly recorded than in the woodcuts and paintings of Japan. Indeed, no inhabited coastline has recorded a disaster to equal the 100-foot tsunami that hit Japan on June 15, 1896, killing 27,000 people and destroying 10,000 homes.

The only U.S. coastlines touched frequently by tsunamis are in Alaska and Hawaii. Like coastal residents everywhere, those of Hilo, Hawaii, the state's second largest city, flirt willingly with any natural danger. The harbor faces northeast into the trade winds and the prevailing direction of tsunami approach behind the town rises the active Mauna Loa volcano, whose lava flows have several times almost overrun the town. The town itself sits on an old lava flow, and in 1852, 1855, and 1881 new flows stopped at the very edge of the developed area. In 1935 and 1942 only army bombing diverted yet more lava flows. Although advised to build a protecting wall against the lava, the city has not done so.

Hilo has been the site of repeated disasters since record-keeping began in the 1830s. Tsunamis partially destroyed the town in 1837, 1877, and 1923, and many smaller ones had battered the bay during the years between to remind the inhabitants of the dangers they live with. With increased development the tsunami damage has become more expensive and more spectacular. Tsunami damage has become even more frequent, since intense development crowding the shore means that even small tsunamis take their toll from something. The end of World War II and the massive commitment of technology to development invited an epidemic of disasters. In 1946, residents of Hilo had not seen the attack of a major tsunami for twenty-three years. Undoubtedly, this reinforced the tendency to think of these catastrophes as random acts of God or freaks of nature, rather than as expected natural events that accompany the ever-moving plates of the earth's crust. In late March, several thousand miles north of Hawaii, the windswept Aleutian Islands were shaken by an underwater quake. Waves raced south through the Pacific.

One of the best firsthand accounts of the power and suddenness of the 1946 tsunami is given by Francis Shepard, a marine geologist from Scripps Institute who was renting a cottage on Oahu Island when the waves struck:

I was awakened at 6:30 in the morning by a hissing noise that sounded as if hundreds of locomotives were blowing off steam. I looked out in time to see the water lapping up around the edge of our house, rising 14 feet above its normal level. Grabbing my camera instead of clothes, I rushed out to take photographs. My wife and I watched the rapid recession of the water exposing the narrow reef in front of the house and stranding large numbers of fish. A few minutes later, we saw the water build up at the edge of the reef and move shoreward as a great breaking wave. Because the wave looked very threatening, we dashed in back of the house, getting there just in time to hear the water reach the front porch and smash in all of the glass at the front. As the water swept around the house and into the cane field, we saw our refrigerator carried past us and deposited right side up, eggs unbroken.

Water swept down the escape road to our right, leaving us no chance to get out by car. The neighbors' house, which was vacant at the time, had been completely torn apart, but the back portion of our house still stood, thanks probably to the casuarina [ironwood] trees growing along the edge of the berm in front.

As soon as the second wave had started to retreat, we ran along the beach ridge to our left, and then through the cane field by path to the main road, arriving there just ahead of the third wave.

We found quite a group on the road. The house of one family living at the edge of Kawela Bay had been carried bodily into the cane field and dropped without having done much damage. In fact, their breakfast was still cooking when they were set down. Other unoccupied houses did not fare so well. Some of them were swept into a small pond, and others out into the bay by the retreat of the second wave.

We watched three or four more waves come into Kawela Bay at intervals of fifteen minutes. They had steep fronts, looking very much like the tidal bore that comes up the Bay of Fundy. Just before the eighth wave, I decided that the excitement was abating and ran back to the house to try to rescue some effects, particularly necessary as we were in pajamas and raincoats. Just as I arrived, a wave that must have been the

> largest of the set came roaring in, and I had to run for a tree and climb it as the water surged beneath me. I hung on swaying back and forth as the water roared by into the cane field.

On the island of Hilo, even when the first wave had broken, natives were not alarmed. Many ran out onto the exposed coral reef to gather the fish left waterless by the trough of the first wave. Since tsunamis are long-period waves, the trough and next crest were separated by ten to thirty minutes. People felt confident they could beat nature, or perhaps that there was only going to be one wave. The second wave came on even stronger and higher. A wall of water twenty feet high rushed into the bay, covered the seafloor again, and exploded against Hilo's waterfront. The destruction of buildings, roads, and bridges amounted to $26 million in damage, and ninety-six people were dead.

The very formation of the harbor, which makes Hilo an economically important seaport, also subjects it to greater-than-normal risks from tsunamis and storm waves. The ocean floor of the harbor causes waves to converge, then the funnel-shaped harbor concentrates this effect. Even waves from normal storm conditions pound the shore severely, and in 1946 they knocked out a seawall that the tsunamis had left intact though probably weakened. After the April tsunamis, those who tally disaster damage found that no other area in the Hawaiian islands had suffered as spectacular damage as Hilo, even though wave heights in other places had topped fifty feet. The answer is simply that Hilo is particularly vulnerable by virtue of its location and the shape of its harbor.

In 1952, and again in 1957, tsunamis struck, causing $300 thousand and $400 thousand, respectively, in damage. In 1960, more tsunami waves, thirty-five feet high, battered Hilo, leaving $22 million in damage and sixty-one dead.

In 1946, scientists set about inventing tsunami warning

systems. At first they relied on seismographs and underwater sensors. This approach hardly gave enough advance warning of waves traveling 500 miles per hour. Today the Pacific Tsunami Warning System (PTWS) detects tsunamis in the air. The vertical-motion quakes which produce tsunamis also create acoustic waves in the ionosphere. These so-called "ripples," traveling at 7,800 miles per hour, quickly outdistance the tsunamis. Because they disturb the air layer that reflects radio waves, they can be detected electronically. On vulnerable shores local warnings are triggered by special tide and wave gauges. When the instruments feel the unusual wave-energy patterns that precede tsunamis, they automatically sound an alarm.

Through the famous Murphy's Law (anything that can go wrong will go wrong) human nature often negates its own ingenuity. Oceanographer Willard Bascom reports that the first alarm system in Hilo, Hawaii, in the early 1950s, was connected directly to the police chief's office. He ordered that the tide gauge warnings should be ignored until a phone call had been placed to the dock, where a man would confirm or deny the presence of unusual waves. (One would have wondered, if the phone were not answered, whether the dock employee was out to lunch or had disappeared in a wave.) That situation has been corrected now, but in both Hawaii and California the sounding of tsunami warnings has sometimes resulted in traffic jams of the idle and curious, who race down to the shore to watch the big wave, preventing those who live along the shore from making an appropriately rapid retreat.

The only North American coast touched by tsunamis in recorded history is the West Coast, and there, except in Alaska, the waves have been relatively low and infrequent. Northern California and Oregon have occasionally experienced waves of three to ten feet high, and the 1964

Alaskan quake raked these coasts with a series of waves, the highest reaching twelve feet. These coasts have seen higher waves in storms, but the storm waves do not travel as quickly or strike so unexpectedly. At Crescent City, California, the harbor is almost enclosed by a breakwater and two jetties. The structures have generally kept the fishing fleet and lumber docks safe from stormy seas. On Good Friday, 1964, however, tsunamis generated by the Alaskan quake leaped the breakwater and poured across the waterfront. The waves inundated twenty-nine blocks of the waterfront, wrecked twenty-one boats in the snug little harbor, and caused $2.7 million in damage. Experts predict that the biggest tsunami waves striking the city could reach seventeen feet.

The tsunamis which travel the vast reaches of the Pacific are a normal part of that basin's geological activity. Their explosive power at the coast is enhanced because deepwater approaches close to shore and often the underwater shelf is narrow and relatively steep. All this allows the waves to approach close to the beach before meeting resistance. A steep shelf means a rapidly rising breaker. Residents of vulnerable beaches may have some luck in defending themselves for a while against cliff collapse and storms, but retreat is the only defense against tsunamis. And these giant waves in their brief attacks often tear out all other defenses.

Hurricanes and Storms

Somewhere on earth winds are always blowing. Perhaps they blow most consistently along the coasts. Welcoming the winds when we want them and in their usual summer gentleness, we do not like to think that these same winds, or at least the same air, can also destroy us. We look upon violent storms as strangers in a pastoral landscape. Hurricanes aborning in the Caribbean or Atlantic are first called "disturbances." Later we endow them with women's names, and the media encourages us to talk of them as if they were angry harpies bent on wreaking havoc in our peaceful households.

There is no such discontinuity in nature, and even man is hard-pressed to draw exact lines between a strong wind and a storm, a storm and a hurricane. Once a storm or hurricane reaches a certain intensity it has distinct characteristics, but it still obeys the same physical principles as a breeze. If we put the strength of air movements on a graph, most of the time the line would rise and fall within certain limits. But at widely spaced intervals it would rise far above those limits, and we would call those high peaks storms. Those wind peaks or storms create changes on the beach with a suddenness and size far out of proportion to their frequency. The wind itself drives sand with enough force to erode rock, but the wind exercises most of its force through an intermediary—water.

Storm waves are the embodiment of a small but dangerous part of a storm's energy. No better condition exists for breeding waves than a storm at sea. Wind velocity is high. It often blows for several days or even weeks, and its fetch reaches over great distances. (A hurricane, rather compact and fast-moving, is sometimes less devastating than a slow, broad, winter storm.) The amount of energy stored in the 40,000- to 50,000-foot-high cloud walls of a hurricane challenges the mind's capacity. In one day an average storm might release heat energy equivalent to the explosion of two-dozen 20-megaton hydrogen bombs. Or in two days its heat energy could generate enough electricity to supply American electrical demands for almost a year. Even the small fraction of this energy that is actually transferred to the waves of the sea is more than human defenses can cope with except in very limited areas.

Of course we try to cope. In February 1978, when winter storm waves twelve feet high hit southern California, the famous former bellhop Flip Wilson was back carrying bags. This time he was hauling sandbags to protect the foundation of his friends' houses at Malibu Beach. Dozens of actors and actresses issued passionate appeals for help, not unlike the distress cries of the cinema's besieged wagon trains and passengers on sinking ships. A little more slowly than the cavalry, the U. S. Army Corps of Engineers will probably come to the rescue with dredges pumping up new sand. But even if the Malibu people could find protection from twelve-foot waves, their relief is only temporary. The 1978 waves were gentle compared to the forty-foot waves seen on the same beaches not long before the golden age of movies turned a wild shore into a wall of homes. Scientists studying shoreline erosion rates in Southern California have established a cyclical recurrence of big waves similar to the cycles of hurricanes on the East Coast. Big waves will return to destroy most ocean-front houses.

The storm waves, whose height and length store enormous energy, are lifted on the domed back of a "storm surge" rising above the normal sea level. In the center of the storm atmospheric pressure is low. Around the storm the air continues to press down on the water with more normal weight. The water responds by bulging upward in the low pressure. A home experimenter can create a small replica of a storm surge in a saucer of water. Touch the water with your lips, then raise them a fraction of an inch above the surface. Now breathe in fast and hard. Suddenly the water rises and you are drinking from the surge. You did not suck the water in. You lowered air pressure below your lips, and all around the higher pressure pushed down as before. Underneath a large storm a surge on top of a high tide may rise twenty-five feet above normal water levels. Except for New England, very little land within a few miles of the Atlantic and Gulf coasts is more than twenty feet above sea level, and the average sand dunes are five to ten feet high. A major storm surge mounted by violent waves advances on the coast with all the advantage mounted cavalry once enjoyed among foot soldiers. Some 75 per cent of the recorded eighteen thousand who have died on the U.S. coasts in hurricanes were killed by a combination of waves and floods caused by storm surges.

When people watch the hurricane and survey its aftermath, they see a beach strewn with litter, dunes leveled, new inlets cut through entire islands, old inlets closed, and new islands created. The only change so sudden and complete in our life may be death itself. But the beach is far from dead. It has simply responded naturally to a normal and short-lived change in the local energy pattern. In many cases waves washing over the dunes add material to the inland portions of barrier islands and thus make them wider and more long-lived.

None of the four essentials of shoreline equilibrium—material, energy, shape, and sea level—are destroyed. The interplay speeds up, and tradeoffs that sometimes take months or years happen in minutes or hours. The inertia of the materials and the shape of the shore are overwhelmed by the focused energy of an intense storm. The strong winds and waves pick up much more sand than usual, carrying it inland into the dunes or seaward into the waves. Covered by the storm surge, the beach temporarily loses its usual position in the marine environment and becomes an underwater or offshore bar. In this situation it is entirely covered by the surf zone and subject to much stronger longshore currents—the energy trains that are responsible for the transport of beach materials. Under normal conditions these currents and the waves bring in about as much sand as they carry off, but during the storm the boiling surf keeps the sand particles in suspension, and because the energy of the big waves reaches to deeper water, the sand cannot settle until it is either far offshore in quiet depths or until it is washed up on land to reshape the beach.

In the shore environment, equilibrium does not mean stagnation or stability. It means a constant, dynamic balancing of materials, energies, and shapes. The person who is horrified by the destruction of the dunes or the beach immediately after a hurricane is like a baseball fan predicting doom because the opposing team has hit a grand slam the second inning.

Suicide on the Tracks

In silent films and cartoons there is a classic scene in which the heroine is tied to a railroad track and the train is coming. Time and again she escapes. This scene is reenacted in reality by ordinary people, whose numbers are a thousandfold greater than those of the film heroes. The Gulf Coast and the Atlantic Seaboard are the railroad tracks, and the trains are hurricanes and winter storms. Instead of people tied to tracks, there are houses and possessions. The owners, banks, and insurance companies have tied the knots by refusing to allow homeowners to move or abandon damaged homes. Unfortunately, once the storm has come and gone, sympathy focuses our attention on the victim and we forget how she came to be tied on the tracks.

Hurricanes and violent storms along the Gulf and Atlantic coasts are sometimes more regular than American trains, and service is expanding. Over the past seventy years an average of eight tropical cyclones have occurred in this region per year. In the past forty years the average increased to nine; and in the past ten years, to ten.

On many parts of the coast real estate salespeople and homeowners will say they live in a trouble-free area. Most of the residents and some of the salespeople are too new on the coast to know. The 60,000 people who live in the Florida Keys feel they don't have a bad storm problem. Most are not natives, nor are they interested in local history. They do not know that in 1846 a hurricane destroyed Key West while water was five feet deep in its main street. The inhabitants clustered for safety on the highest piece of land —seventeen feet high, with waves threatening to push them off. Few people in the Keys know of the hurricane-battered ship that threw out a desperate but successful anchor in 1796. When the storm moved on, the ship was stranded in the middle of Elliot Key in Biscayne Bay. The boat was burned to salvage the iron fittings. Even 1935 is forgotten. That year a storm sent eighteen feet of water over the Keys, destroyed a railroad being built where the escape road now stands, and killed 400 people.

What the 60,000 people in the Keys don't know may one day kill thousands of them. They have only a two-lane road to the mainland. It passes over sixty bridges, several always in poor repair. A single accident could choke the road, and in the best times only 700 cars per hour can pass along the road. Multiply 700 by the maximum 12-hour lead the Hurricane Center can provide, and you have 8,400 escapees and over 50,000 victims.

The feeling of security that has brought so many people

to the Keys (20,000 in mobile homes) presently prevails on most of the coastline. Florida has not had a major storm since 1950. The Southeast has been blessed since 1954. The Northeast has also escaped since the 1950s. Are we playing with the definition of "major storms," reserving it for infrequent monsters? The National Hurricane Center in Miami rates the weakest hurricanes 1 and the strongest 5. Agnes in 1972 rated 1. Only Camille in 1969 and the 1935 Labor Day storm along the central Atlantic Coast have rated a 5 in this century. A major storm is a 3, 4, or 5. As if to tempt development, these big storms seem to visit in cycles. In the 1940s they crossed the Florida panhandle. In the 1950s they roared up the East Coast. In the 1960s and 1970s all eight major storms struck the Gulf Coast. These cycles induce many regions to consider themselves safe, but the broader perspective of statistics says any homeowner on the beach is gambling against overwhelming odds.

The history of the coastline is a history of storms. Marine geologists who study the continental shelf and beaches work in the framework of geologic time. Studying oceanographic processes over millions of years, any one year is but a point in time, and the years between storm cycles shrink into nothing. Severe storms are so regular that they may be a continuous tensing and relaxing of the atmospheric muscle. Many geologists believe that storm conditions are completely responsible for the distribution of shelf sediments which create and nourish the beaches.

Studies of Atlantic Coast storms over the forty-year period from 1921 to 1962 show 170 damaging storms, most of them occurring in September, October, and November. The same studies produced the following figures showing the likely period of time that will elapse between damaging storms on a given coast.

State	*Years likely between storms*
Maine	1.6
New Hampshire	1.9
Massachusetts*	0.8
Rhode Island	1.5
Connecticut	1.6
New York	1.4
New Jersey	1.4
Delaware	3.0
Maryland	2.8
Virginia	1.8
North Carolina	1.5
South Carolina	2.8
Georgia	4.2
Florida	1.6

But this table is misleading. It shows the state with least developed shoreline—Georgia—as the safest. Yet, an 1893 storm brought a twenty-foot surge to Savannah and killed two thousand people. This same storm surge covered Hiltonhead in South Carolina, where a salesman who grew up in the Midwest told a worried prospect there would never be reason to evacuate. His prospect never bought a beachfront lot there or anywhere else. He was Dr. Neil Frank, Director of the National Hurricane Center.

It is perhaps some measure of human hope in perfectibility that allows even frequent natural disasters to be dismissed as acts of God. In 1879 some residents of Beaufort, North Carolina, attributed a devastating hurricane to

* The Massachusetts figures are misleading, since the northern Massachusetts coast is particularly susceptible to northeasters while the southern coast receives the impact of southerlies. Taken by themselves, each section is roughly equivalent to Rhode Island or New Hampshire.

an act of God directed against certain neighbors who had been dancing on Sunday nights at the local hotel.

The modern coastal resident seems little wiser about the reasons for storm damage. A year after the great Ash Wednesday storm of March 1962 (not a hurricane) did $192 million damage along the East Coast, geographers from Clark and Toronto universities studied and interviewed owners of beach property. They found, not unexpectedly, that these people had above-average incomes and educations. Property owners were not innocent sitting pigeons set up by unscrupulous real estate agents and developers. Two thirds of them said that when they had bought their land they were aware of the history of destructive storms along the coast. At the time of the interview 90 per cent had already experienced storms and about 50 per cent had suffered water damage, but they intended to stay on. Contrary to the general tone of media coverage, it is simply untrue that storms are taking their toll on unsuspecting and innocent victims.

"Men cannot build houses upon sand and expect them to stand now any more than they could in olden times . . . Summer seaside resorts must be built far enough above the tide line to insure safety as well as patronage. People are wary of making hairbreadth escapes in seeking health and rest." So said the Raleigh, North Carolina, *Observer* on August 20, 1870. Since then, thousands of citizens in North Carolina and other coastal states have made hairbreadth escapes. And many have failed. And with each new storm have come new visions of a trouble-free time.

In 1938 a Federal Writer's Project report said of that year's major hurricane: "There are earnest proposals that seaside resorts pass zoning laws. The New England Council hopes to persuade owners to build cottages further inland instead of at the shore's edge . . . Errors of centuries of haphazard building may now be rectified." In 1954, and again in 1962, storms destroyed buildings on the exact sites where earlier buildings had disappeared in the 1938 storms.

One reason history repeats itself year after year is because the media reflects an ironic syndrome that is so often fatal to beach property owners themselves. No matter how often nature demonstrates that coastal storms are the normal pattern of events that moves the beaches, property owners tell themselves there is no escape. Rather than get off the tracks, they stubbornly blind themselves to the fact that the next train will come sooner or later along their track. Sympathetic reporters and television and radio commentators echo the landowners' dismay and praise their stubbornness.

Although 90 per cent of the people interviewed by the Toronto geographers had experienced storms personally, only two thirds expected to experience more storms. And of those who had not only been through the storm but suffered damage, some 50 per cent believed that was the last of the storms.

Property owners clearly responded with two kinds of reasoning. One group comforted themselves with finding a definite order in the natural events. They saw either a cyclical pattern in the visitation of the storms, or a unique set of causal circumstances unlikely to happen again. Their reasoning was based on folklore and very sparse and disorganized observations. But it helped them predict a comfortable future and allowed them to return to unworried enjoyment of their property. The second group did not dare to rush in where professional scientists tread lightly. In fact, they denied that any human could predict the future course of nature on the beach. One man said, "God doesn't tell us things like that."

These two groups differ widely in their reasoning but not

in their direction. They both refuse to recognize human vulnerability in the face of nature's most powerful forces. Compared to primitive people's consistent respect for nature, modern men and women show remarkable lack of perspective. Primitive mythology and religion saw storms as acts of the gods, but the gods were nature and ever present. Nature was an endless drama, with the major characters always reappearing. But when we call a storm an act of God, we mean it is the sudden and inscrutable display of power sent from some world beyond the here and now, something that is not ours in the way that we think of pastoral nature—the birds, sunlight, trees, fields, cows, and dogs—being ours.

Almost no one blames himself for the destruction done by a storm. The most common reaction is to turn to the government for protection or restitution. Property owners in the Toronto study commonly turned to such technological solutions as seawalls or dune-building programs. Just as early settlers called on the cavalry to provide technological defenses against "vicious" Indians, so beach dwellers call on the engineers of the modern Army to fight storms. This traditional American stubbornness in the face of adversity settled and populated the continent with Western civilization, but it is destroying the edge of the continent. Seldom is the response to a storm a call for land-use planning or zoning. And in modern times there has seldom been a retreat from an untenable position.

Our intransigence has never been more clearly demonstrated than in the case of Hurricane Camille, which struck the Gulf Coast of Mississippi and Louisiana on August 17, 1969. Hardest hit was a sixty-mile-wide, three-county area of Mississippi that contained the cities of Long Beach, Pass Christian, Bay St. Louis, Waveland, and Gulfport. In poverty-stricken Mississippi, as elsewhere around the

United States, the coast, or more particularly land within sight of water, is the home of affluent middle- and upper-class citizens. In the decade before Camille, the area had a 26.9 per cent growth in population while the rest of Mississippi had stood still or lost population, and the nation had grown by 12 per cent. Industry was coming in and choice land was going fast. Almost a quarter of a million people were living on a strip of coast some two miles deep, a density of almost two thousand people per square mile. The property within two or three blocks of the beaches was almost entirely bought up by white-collar workers with good educations and with sufficient incomes to buy and maintain large homes.

In this warm climate the beaches are usable most of the year and the breezes are generally welcome. Despite destructive storms and extensive flooding in 1947 and 1965 (Betsy), the demand for property near the beach has never slackened. On the "Gold Coast" of the Pass Christian–Gulfport–Long Beach area, almost all the flood-prone land is occupied by private housing, motels, and shopping centers.

In 1969 Camille approached in much the same way hurricanes have always approached this coast. On August 14 she was a tropical storm near Grand Cayman Island in the Caribbean. The next day, over Cuba, she reached hurricane intensity. On the sixteenth she stalled in the Gulf of Mexico and intensified. Winds were now 150 miles per hour and the storm began to move north–northwest. Military planes scouted the storm while it was still out in the Gulf and reported it to be a small storm but one of the most powerful ever to enter the region. A barometric pressure of 26.61 was the lowest ever recorded in the Western Hemisphere and ideal for the creation of a storm surge. At 10 P.M. the storm crossed the coast near Waveland with winds estimated from 125 to 200 miles per hour. (The most accurate and detailed of recorded observations did not verify winds above 125 miles per hour.) Because of its timing, size, and intensity, the storm brought with it a great dome of water superimposed on high tides. The surge varied from 10 feet at Pascagoula to 21.4 feet at Pass Christian. Across the top of the surge raced large storm waves.

In Pass Christian twenty-five people in the new three-story Richelieu Apartments decorated the third floor for a hurricane party. The celebrants welcomed the police chief and laughed as he took down the names of their next of kin in a polite southern attempt to warn them that they should join 70,000 others fleeing the storm. A short while later, a woman resting in her room heard her picture window blow out. Waves began to beat into the apartment. Clutching part of her sofa, she floated out into the surf. Soon the tide had dragged her so far offshore she could not see land. After some ten hours a new high tide brought her to the beach almost five miles from the apartments. A ten-year-old boy also survived. The other twenty-three were taken care of by their next of kin.

The Richelieu Apartments party was reminiscent of another hurricane party held in the Hotel Dernieres in New Orleans over a century ago. That hotel also disappeared. A hurricane like Camille hitting New Orleans today would force 500,000 to evacuate if the traffic jams allowed. Fortunately, after some meandering, Camille spared New Orleans and passed inland. In its wake, 2,822 homes disappeared and almost 40,000 more were damaged, mainly by rising water; 22,000 families needed recovery assistance; 533 small businesses were damaged or destroyed; 80 per cent of the area's motels were wrecked; 4,310 people were injured or ill, and 130 had died.

For a while the storm seemed to have instilled in residents some respect for nature and some sense of man's fragility. After the storm, the staff of the *Sea Coast Echo* in Bay St. Louis printed a booklet on the storm. The introduction ended with these words: "American flags were never so cherished. Lives never held in more reverence."

The *Echo*'s pronouncement was probably true—but only temporarily. The residents of the Mississippi coast differ little from Americans elsewhere in their inability to balance immediate comfort against risks whose appearance is certain but unpredictable. The greatest citizen response to Camille came immediately after the storm during recovery efforts. These efforts yielded visible results, and perhaps the miraculous clean-up wrought with the aid of massive technology and economic transfusions restored man's confidence in his ability to confront nature. Today the Gulf Coast of Mississippi continues to grow rapidly. The beachfront is once more filled with houses, motels, and restaurants. Lots with a view of the Gulf sell for $20,000 to $30,000. Some of the victims of Camille have moved inland and newcomers have taken their places, but it is hardly likely that the newcomers have never heard of storms Betsy and Camille.

What these disasters have cost the taxpayer in emergency aid is not easily figured. Over $65 million was spent in Mississippi in restoring public facilities, and considerably more was spent through the efforts of the U. S. Army Corps of Engineers, loan guarantees, and subsidized flood insurance. Local taxpayers also suffered from their own tolerance toward building in high-risk areas. Schools survived only through federal assistance. Everywhere along the coast physical erosion of the beach was accompanied by serious economic erosion of the tax base. In Pass Christian, 50 per cent of the town's taxable property value was destroyed. The figure in Long Beach reached 80 per cent.

Not all of the damage of Camille could have been avoided or moderated. We are not that clever, nor is a storm that simple. Some risks have to be taken. In the final analysis, we live and die by nature in spite of our aloofness, but we could be more cautious about the way we do it.

Much of the $1.3 billion damage to the Gulf Coast could have been avoided, and future damage could still be avoided by wise planning and shoreline conservation practices. But why go to all the trouble, since the U. S. Geologic Survey lists Camille as a 170-year storm?—that is to say, since the chance of this particular combination of winds, waves, tides, and air pressures occurring simultaneously in any given year is about 1 in 170. (The great storm surge of 1953 that breached the dikes in Holland was an 800-year storm.) Since the storm was such a statistical rarity, and since the only type of shoreline structure that could have survived the direct onslaught would have been a waterproofed and reinforced concrete pillbox, we tend to chalk it off to bad luck or an act of God and go on about our business.

Such security is a delusion. Each and every year there is still 1 chance in 170 that a similar storm will occur. And even Betsy, a smaller storm, wrought $400 million damage only four years earlier. In Waveland, homes destroyed by Betsy and rebuilt in the same way on the same site were redestroyed by Camille.

We can understand the odds better if we think of the 100-year flood or storm. (Most building codes and flood-area zoning are based on this frequency.) If you build for a 100-year storm, that means that each and every year you are taking a 1-in-100 chance. Over a period of 50 years the chance is almost 50–50 that you will be wiped out. In other words, your 50-year gamble has the same odds as flipping a coin. Your 25-year gamble has about a 1-in-4 chance.

On the hurricane coasts all odds are shaky, and since we

do not have to take them, the wisest course is not to bet at all. Many of the Gulf Coast's earlier residents chose never to bet, though they knew even less about the certainty of storms than we do. Jefferson Davis, for example, built his house adjacent to the shore, and, enjoying a combination of high elevation and sound construction, it survived Camille. Those who settled on the shoreline long ago generally avoided low areas. Modern houses which intentionally or unintentionally were built on high land elevations, survived Camille without fundamental damage. Houses at lower elevations were destroyed. Each row of houses furnished material that acted as battering rams for the next row of houses.

Yet citizens alone are not at fault. They are often locked into an economic and social system that will not let them seek alternatives. Most homes on the Gulf Coast are insured and mortgaged. When owners try to collect on their claims, they are faced with this proposition: either forfeit the money to the bank to pay off the mortgage, or rebuild; and even if you choose to rebuild, you cannot receive any money until construction is finished, since insurance checks are made jointly to the bank and owners. Only when the bank has forced the homeowner at his own expense to resink his anchor in the beach sand, do the banks ease his burden. On its face this seems equitable. In practice it is absurd. Not only does the system encourage rebuilding in the same flimsy fashion, especially if the entire structure has not been washed away in the storm, but it tends to duplicate disaster after disaster. Understandably, banks want to collect on their mortgages or restore lost collateral; yet, at the same time, it is not in the bank's interest to ignore risks to the borrower, who must rebuild in the danger zone, or to the taxpayer, who will once again have to foot the bill for emergency relief.

With two or three hurricanes crossing the Gulf and Atlantic coasts each year, the lessons of Camille are applicable from Texas to Maine.

Lesson 1

Don't build on Oceanfront Avenue. The only safe structure to build on a beach is a reinforced concrete pillbox, and even that can be undermined.

The shoreline dunes must remain an undisturbed storm buffer. In heavily forested delta coasts the trees must also be maintained as buffers to break wind and wave energy and to act as debris jams. Study the storm history and beach dynamics to determine how wide a buffer strip is needed. These vary according to locale. Arbitrary or politically set dune widths such as sixteen feet at Bogue Bank, North Carolina, are simply self-indulgent when we consider that Hurricane Hazel ripped away seventy-five feet of Bogue Bank in 1954. Miles Hayes, a beach expert at the University of South Carolina, measured uniform beach erosion of eight hundred feet where Hurricane Carla struck the Matagorda Peninsula in Texas in 1961.

Lesson 2

Avoid nearshore construction at low elevations. On the Mississippi coast this means any elevations below fifteen feet. Danger-zone figures are available from the U. S. Army Corps of Engineers for virtually every beach in the United States. Limiting construction will mean, of course, fewer beachfront structures, but the wisdom of this limitation has been demonstrated repeatedly. Some homeowners try to circumvent the elevation problem by building homes on piers with sacrificial first floors or with no grade-level dwelling areas or walls. Unfortunately, they often forget to attach their houses firmly to the posts. In other cases, ordinary storm debris knocks out the supporting poles. The January 1973 storm along North Carolina's Outer Banks undermined even the pilings that houses were built on; in some places so much sand accumulated beneath the houses that the waves, unable to wash underneath, broke against the walls, windows, and doors, filling the houses with water and sand or eventually tearing them apart.

Houses attached to piers or wooden poles survive until flooding, but Camille and other storms have demonstrated that these supports are often tragically flimsy. Because of infrequent points of attachment to their foundations, houses are easily lifted from poles and moved at will by waves and floods. One such house in Waveland, Mississippi, was detached from poles ten feet high and its fragments deposited on the opposite side of a marsh a quarter of a mile away in pine branches twenty-five to fifty-feet high. Its poles showed no evidence of damage. The entire house should have been attached by metal bands to poles, and the poles should have extended into and been part of the framing system.

Lesson 3

Only strict building codes can provide even moderate certainty that houses will not crumble in the first full storm. Traditional wood-frame practices can be adapted for coastal conditions without boosting cost more than 5 or 10 per cent. In 1971, however, the citizens of Hancock County, Mississippi, the area hardest hit by Camille, turned down by referendum a proposed building code for all new construction in the area.

In hurricane-prone areas, as in earthquake zones, structural continuity is the key to safe construction. Continuity simply means that all parts are securely fastened or tied together. Elevated water tanks are good examples of continuous construction, and not one of these went down during Camille despite their height and the fact that they often served as log jams for debris from local buildings. Buildings supported by continuous rigid-steel frames remained standing but were often stripped skeletons—the loosely attached brick or wood coverings ripped off by water and wind. The worst disasters of all involved concrete block structures, which for all practical purposes are put together like building blocks. Mortar keeps blocks in place but it does not glue the blocks to each other, as many laymen believe. The cement simply penetrates a short distance into the pores of the block and creates a form-fitting seat. Wood-frame houses survive much better than block houses, although they sometimes float away. It is a toss-up whether to take a chance on recovering a house that floated away (as some people did after Camille) or risk the destruction of a wooden house securely anchored to a concrete slab.

Even houses which retain their sites often lose their roofs. Many victims of Camille were left with four walls and no roof, because the roof trusses were attached to the walls only by nails driven at an angle through the rafters and joists. Roofs should be constructed of trusses fastened to the walls with special hurricane clips.

Lesson 3 is of absolutely no use if Lessons 1 and 2 are ignored. No waterfront homes survive a major storm center.

Lesson 4

Weather forecasters are not God. Although everyone talks about how fallible they are, most of us still choose to believe them when a crisis occurs. Others say the Hurricane Center exaggerates. Forecasters are caught in a dilemma. Twelve hours is the best lead time they can give, and even at that late hour a 40- to 50-mile-wide storm center can wander 50 miles off the mark. So for every twelve-hour warning, the population along 150 miles of coast are advised to evacuate. Twenty-four hours before the storm the margin of error could mean warning people 100 miles on either side of the storm. This safety margin has encouraged many coastal residents to say that the Hurricane Center often cries wolf.

Satellites, radar, and observation planes are the forecaster's eyes and ears, and they are, of course, better than guesswork. However, storms are unpredictable even for the meteorologist. The weather forecaster is not in the driver's seat. Assume that any nearby storm will hit your area directly. Seek the best shelter you can find.

Lesson 5

Choose your insurance company carefully. Because of the complexity of forces at work during a hurricane on the coast, it is often difficult to prove what ruined your house. Many companies, including some of the largest, and best-known, use the uncertain origin of damages to avoid paying claims. In one instance in Waveland, Mississippi, after Camille, two adjacent houses of similar description received almost identical flood damage. One homeowner received over $4,000 in payments, while the other received only $150. One company decided that rising water did the damage (which was not covered); the other concluded that the damage resulted from falling trees and rainwater. Before buying insurance, talk to your friends and neighbors. You will quickly discover there *is* a difference.

Lesson 6

Because you have survived a hurricane, don't think you have seen the worst. While a hurricane's winds cover a 200- to 300-mile front, the real fury is only in the 40- to 50-mile-wide center. Many coastal residents ride out the fringes of a storm and become "expert survivors" who ignore the warnings next time. Almost 80 per cent of the present coastal residents have not been through a major storm, and even fewer have survived the center of a storm.

Prospective coastal homeowners who have learned the lessons of Camille and other storms will increase the survival chances of their house and family. Nor will they become a burden on the taxpayers, who provide emergency aid, loans, and insurance guarantees. Along both Gulf and Atlantic coasts people are ignoring all of these lessons, but there is no doubt that construction codes in the Mississippi region have improved. Two major hurricanes in four years have made an impact; but, incredibly enough, the impact exists more on paper than in fact.

Meanwhile the Atlantic Coast has prospered for twenty years without a major storm. The last major storm to hit the North Miami area came in 1950. Since then, some 3.5 million people have moved in. Along the central Atlantic Coast some 19 million people have moved to the shore, and 96 per cent of them have never seen a big hurricane. Taking advantage of this prolonged sunny day, ignorant and unethical developers are selling lots and homes in unprecedented quantities. Some of the buyers are ignorant. However, most people are simply practicing the economics of joy. They ignore the 18,000 Americans killed and billions of dollars in property loss caused by storms since 1900.

Perhaps by the time these pages reach print there will have been one or more "natural disasters" on the coast. Dr. Neil Frank, of the Hurricane Center, says he can tick off half a dozen places on the East Coast where a major storm could kill 10,000 people in one day. Whenever the disasters occur, the Red Cross, the National Guard, and the Salvation Army will come to help; the government will declare a disaster; and the Vice-President or President will arrive by helicopter to express his shock, surprise, and dismay with the stock words: "This is one of the worst natural disasters in the history of our country. This administration will do all it can to help the brave people of ——————."

Atlantic and Gulf Coast Hurricanes

1635 August hurricane in New England with 20-foot tides.

1796 Storm surge submerges most of Florida Keys.

1815 September hurricane tops all dunes on Long Island's western beaches.

1831 Fishing village on Grand Isle, Louisiana, destroyed and 150 people killed.

1842 Hurricane at Cedar Key, Florida, brings 20-foot surge.

1843 Hurricane destroys Port Leon, Florida. Town never rebuilt.

1844 Port St. Joe, Florida, destroyed.

1846 Hurricane destroys Key West. Water 5 feet deep in Main Street.

1848 Hurricane with 15-foot surge floods Tampa area.

1886 Hurricane destroys thriving Texas seaport of Indianola and all residents.

1893 Hurricane with 20-foot storm surge floods Hiltonhead, South Carolina; kills 2,000 people in Savannah, Georgia.

1893 Hurricane kills 1,150 people in Cheniere Caminada, in Louisiana.

1900 City of Galveston destroyed and 6,000 people killed.

1909 Hurricane kills 350 in Mississippi–Louisiana coast. 15-foot tides.

1919 Hurricane traveling toward Texas kills 300 at Key West, 500 at Corpus Christi. $22.7 million damage, 16-foot tides.

1926 243 people killed in Miami area by 13-foot tides. Development of Miami Beach stopped. $112 million damage.

1928 1,836 killed in Florida. $76 million damage.

1935 "Labor Day" storm kills 408 in Florida Keys with 20-foot tides. $46 million in damage.

1938 Late summer storm kills 600 in New England and on Long Island. $400 million damage by 25-foot storm surge.

1944 Hurricane kills 600 on Atlantic Coast and in New England. $400 million in damage.

1954 Hurricane Carol kills 60 people in Rhode Island and leaves $461 million damage.

1962 March storm from north does $192 million damage on Atlantic Coast.

1969 Hurricane Camille kills 130 coastal residents with 25-foot surge. 2,822 homes destroyed; 40,000 homes damaged; 4,310 people injured. Over $1 billion damage.

8

The Will to Power: An Old World Heritage

THE STORY OF mankind's attempt to live with the moving beaches begins on the edge of a sea that once lay dry and scorched. Ten million years ago between the continents of Europe and Africa the floor of the earth plunged downward ten thousand feet below sea level into an enormous gaping valley that is now filled with the Mediterranean Sea. There was no water then, only a parched and lifeless desert many times more desolate and thousands of times larger than Death Valley. Rivers from Africa, Europe, and Asia Minor poured into the valley only to be boiled away by the sun faster than it could accumulate. The Atlantic Ocean was forbidden entrance by a blockage across the Straits of Gibraltar.

The Phoenicians, Egyptians, Hebrews, Greeks, Etruscans, and Romans always knew the Mediterranean as a sea full of motion and life. The ocean had long before broken through the Straits of Gibraltar to fill the great desert and help create the climate and trade routes that nourished civilization in Europe and the Middle East.

As soon as oared galleys and graceful triremes began to

sail the Mediterranean, people began to worry about the shores to which these vessels sailed. While the boats often set out from natural harbors or from the banks of broad quiet rivers, the stops along their trade routes afforded few such havens. The most common method of landing was to ground the boat as close to shore as possible. Early Greek ships accommodated themselves to natural conditions by carrying two ladders—a short one for steeply sloping (coarse-grained) beaches where the boat touched bottom close to the waterline, and a long one for the more gently shoaling (fine-grained) beaches where the boat had to remain some distance from shore. Once unburdened of its human cargo, the boat itself might be pulled ashore. As trade began to involve larger boats, heavier cargoes, and a larger number of hostile ships and pirates, these simple landing arrangements became cumbersome. A number of ancient "sea" victories were won when a victim was caught with his boat beached.

Pirates, rival navies, heavy ships, intensified trading, and growing nationalism all led inevitably toward modification of the shoreline. Early Greek harbors were built away from the great cities to assure neutrality of commerce and to protect the cities from both military enemies and natural disaster. Today's confused travelers to Athens, who think they have been mistakenly shunted away from their destination because their tickets are marked Piraeus, are simply following the ancient trade routes. The separation of port and city allowed builders to choose port sites where there were natural harbors.

Whenever the Greeks settled in the Mediterranean world they chose harbors that nature had designed and maintained. Their engineering modifications were few and modest. Today's most famous Mediterranean ports were first internationalized by the Greek settlers or, in some cases, by the Phoenicians before them. These ports include Marseilles, Naples, Barcelona, and Trieste.

While these lessons in living with nature survive, our tradition of coastal management derives not from the founders of Piraeus but from the engineers whose ports disappeared long ago. Two thousand years ago, just as today, growth, wealth, and power tempted people toward bolder and bolder reshaping of the natural environment. Hellenistic Greece began to protect its harbors with stone breakwaters or moles. And as commerce grew in importance, the ports began to be built within city walls. Two of the Seven Wonders of the Ancient World are Hellenic harbors—one at Rhodes where ships passed beneath the 120-foot-high Colossus, and another at Alexandria where the world's greatest lighthouse stood. Yet Greek accomplishments are simple compared to the works of their conquerors.

As governments became increasingly powerful—both technologically and militarily—they came more and more to rely on their own engineering. The *Oxford Classical Dictionary* declares that the Romans "gained complete independence of nature." This is what the Romans wanted, but the admiring dictionary essayist overlooks the evidence on the Roman coast. Roman engineering was greater in its mechanical genius than in its broader understanding of natural processes.

In early times goods destined for the mother city of Rome were set ashore some one hundred miles away in the natural harbor of Pozzuoli (Bay of Naples), first used between 1000 and 500 B.C. by Greek settlers. Then in the third century B.C. the Romans began using the port of Ostia at the mouth of the Tiber River only fourteen miles away. This was the most convenient harbor on the low, sandy, central coast of Italy, but it could not serve the purposes of a great empire without drastic alterations. Southwest winds often imperiled ships anchored in the harbor, so much so that a respected guild of *urinatores*, or divers, maintained themselves by salvaging ships and cargoes. The same southwestern exposure, joined with deposition from longshore currents and river-borne silt, repeatedly threatened to close the harbor to shipping. The Romans at first met the challenge with constant dredging. In A.D. 42 the Emperor Claudius sought to outflank the problem by excavating a 150-acre man-made harbor in the marshy plains two and a half miles up the coast from Ostia. He enclosed his harbor in the arms of two enormous breakwaters, each 1,900 feet long and 180 feet wide. After freighters unloaded in the port, ox-drawn barges carried their cargo along a canal that connected the harbor to the Tiber River. Thence they were drawn up the river to the great docks of Rome.

By A.D. 103, however, the Emperor Trajan found Ostia again rendered useless by silt. Behind the Claudian harbor, which could now handle only shallow-draft vessels, Trajan built an entirely artificial hexagonal harbor reached by a channel through the old harbor and connected to the Tiber by an enlarged canal. Trajan's harbor was entirely encased in stone walls and flanked by warehouses, theaters, palaces, markets, and public buildings—a mini-capital by the sea and an engineering marvel.

Yet the fate of Ostia had been sealed years before the work of Claudius and Trajan—sealed by earlier Roman engineers. In the first century B.C., in order to improve the Tiber River channel for shipping, the engineers had utilized an ingenious form of stream channelization. Far upstream they encased the Tiber within steplike walls so that at low water the river was confined to a narrow channel rather than spreading wide and lazy across its broad natural bed. The narrow channel also meant that the river continued to flow fast and scour the bottom of sediment. This relieved the engineers of the once annoying dredging problems. At high water, however, the river no longer overflowed its banks to deposit rich sediment on the surrounding plain. Instead, the sediment was steadily transported to the coast, where longshore currents swept it down to the port of Ostia. No matter what Claudius or Trajan did, siltation remained a constant plague to all who tried to use the port. Today Ostia is two miles inland.

In Tunisia, which was once a part of the Roman Empire, the ancient ports of Carthage and Utica provide a telling contrast between Rome's "independence" from nature and another people's co-operation with nature. Utica was a major Roman port, situated conveniently (for the generals and merchants) at the mouth of the country's only perennial stream, the Mjerda River. Even in the first century B.C. when the city was thriving, the harbor was rapidly silting up. Today from atop the highest ruin in Utica a tourist cannot find the sea some twelve miles away. The older harbor at the Phoenician city of Carthage, on the other hand, was a better natural harbor, though less convenient to the river. The Phoenicians preferred it to Utica, and for good reason. Although the Phoenicians had excavated extensively to make the harbor workable, they chose a site far from the silting effects of the river. The Romans, in destroying the city, blocked the harbor but later reopened it. Even these great engineers recognized a superb natural harbor. The harbor remained in good shape until fifty years ago when the French filled most of it. Nevertheless, Carthage Harbor is still usable by small boats, and a little dredging would quickly improve the abused facility.

In this ancient lesson is a principle that might still guide us today: complex engineering projects often impede progress more than they help. The lesson is more relevant if we add the final detail to our story of classical harbors. In modern Carthage a jetty has been constructed for the presidential palace a short distance down the coast from the old Phoenician harbor. The jetty has stopped the flow of sand along the beach. The ensuing erosion is eating its way steadily toward the ancient harbor. The shoreline at Carthage, however, is a rocky one and is in no immediate danger.

Since the days of the Romans the size and frequency of shoreline disasters have grown in proportion to the energy resources of the country owning the shoreline. History has repeated itself on the shore as regularly as on any military or political battlefield.

Those who hold that the sea can or will be conquered look to the Dutch for proof. American engineering journals are full of the details of the Dutch dike system. And from the pages of *National Geographic* and countless schoolbooks Holland emerges as a thriving modern industrial nation whose cities are surrounded by fertile fields of pasture, grain, vegetables, and of course tulips, all made possible by engineering feats. A dike, we have been led to believe, is a simple wall that can be temporarily mended by plugging a leak with your thumb.

Perhaps nowhere in the world, however, is human settlement so much in danger from the sea as in Holland. With 740 people per square mile, two fifths of its land below sea level, and geological history working to flood its coastline, lowland Holland stands in constant threat of inundation.

On the night of January 31, 1953, disaster did strike. A familiar North Sea storm was blowing—nothing to get excited about. Centuries of dike building had given people faith that these normal stresses could be rebuffed by the tried and tested engineering work of the Netherlands Waterways Authority. Besides, the Dutch had seen flooding before when the Germans breached the dikes on abandoning Holland near the end of World War II. But few realized the wartime flooding had been controlled by limited and selective breaches and the raising of floodgates. The German goal had been simply to render the ground too soggy for pursuit. Even fewer Dutch realized that the spring tide and storm arriving together in 1953 were a very unusual event, at least in the human scale of time—a storm that statistically occurs only once every eight hundred years.

That night the waves battered their way through the dikes in 500 places, 67 of them major breaks. Twenty-six thousand dwellings were toppled or smashed, 600,000 acres of land were flooded, and 1,835 people were killed. The toll would have been much worse except for an elaborate warning system of sirens, bells, phones, messengers, and town criers that reached all but 7 per cent of the population.

Perhaps more significant was the fact that up until the very moment the flood waters broke, many people didn't believe the dikes could fail. Fully 15 per cent of the survivors said that upon hearing the alarm they assumed a holocaust or war had broken out. Several police chiefs and mayors ignored the warnings until the water was already inundating farms and villages.

All this aside, the Dutch, spurred on rather than chastened by the 1953 storm, decided in 1958 to dam up the marshes and estuaries of the southwest coast once and for all. Unlike the Zuider Zee project to the north, very little land reclamation was involved. The billion-dollar Delta project creates four giant dams of concrete, the largest over five miles long, that seal off the mouths of estuaries from the North Sea. The dams effectively straighten out the low irregular coast of the region called Zeeland. Over four hundred miles of seacoast will become the banks of quiet, often stagnant lakes. The big dams will eliminate the need for hundreds of miles of dikes, says the Dutch Government. Its critics argue that once the dikes fall into disrepair, a single breach in the dam wall would flood all of Zeeland. In wartime such a breach would be easier than destroying the many dikes. Meanwhile the engineers are calling the Delta Project the "defeat of the North Sea." Undoubtedly Claudius and Trajan thought they had defeated the Mediterranean.

Whether or not the Dutch standoff is more durable remains to be seen. That approach to coastal problems excites Americans, who accept its finality unquestioningly.

Perhaps in a small crowded nation there is little room to pick and choose the method or scene for development. The engineering of Holland's coastline protects a large population, considerable agriculture, and valuable industry. The Dutch find themselves, therefore, with no choice but to build dikes. Their attempts to protect themselves with engineering work are older than the invention of concrete. Holland had committed itself to a confrontation with the sea long before geology was a science and when nature itself was less threatening. This history of technological commitment should have the stature of prophecy for other nations on the verge of trusting their shoreline to engineering science.

Some four thousand years ago the rapid sea-level rise caused by the still retreating glaciers suddenly slowed. Gentle waves pushed more sand toward the islands' outer edges than could be eroded by storms, the coast began to build seaward. About twenty-eight hundred years ago, however, when sea level in Holland was still eight feet lower than at present, the shoreline began to retreat again.

Right there, just as Western civilization was budding in

Europe, Holland's supply of sand began to run short. The sweeping action of the waves, which had brought the sand from the continental shelf to extend the coast, had also deepened the shelf. As the shelf deepened it became more difficult for wave action to stir its sands and move them shoreward. At the same time the waves, approaching over a steeper shelf, spent less of their energy on a gradually sloping sea bottom and more on the shoreline itself. If natural history were to run its course, Holland would shrink in size as its coast eroded. But human history is moving in a different direction.

Even before the Roman invasion, many Dutch had settled in the lowland where they could take advantage of the fishing in the quiet lagoons behind the barrier islands or in the estuaries of the Rhine and Meuse rivers. Among the curiosities the Romans found as they invaded the "barbarian" lands were the lowland Dutch living on artificial mounds of earth called *terpen*.

The primitive engineering of the terpen sufficed at a time when the surrounding lowlands were still a little above sea level. As sea level rose, however, major floods became an increasing hazard. The historical records shows them occurring in 1374, 1376, 1377, 1379, 1393, 1396, 1405, 1421, 1530, 1531, and 1532. With an existing, well-established population threatened, and even then with no place to go, the Dutch began building dikes.

Holland has continued to expand its size and population, reflecting its importance in European history. Today she stands as a clear example of human and natural history diverging so rapidly that an entire nation must live in dangerous confrontation with the most powerful forces of nature, relying on ever more expensive and more vulnerable technology whose failure could bring unimaginable disaster.

But the most telling story of the Dutch battle with the sea lies not in its famous engineering but in its treatment of the natural beaches that make up most of the shoreline (a story never included in American schoolbooks). Here history reveals the true quality of that country's understanding of coastal problems.

Where the Dutch coast is protected by dunes and barrier islands the Dutch display a profound respect for nature's own devices. They see the beaches not primarily as playgrounds or even economic resources but as a lifebelt around the mainland. The dune ridges and the plants which hold their sand are almost sacred, protected by laws which forbid even walking upon them except for valid scientific purposes.

Despite their enormous engineering feats, the Dutch are aware of the limitations of human ingenuity and resources. Perhaps the worth of any applied science ultimately must be measured not by its own standards but by its practioners' willingness to recognize when it is not needed. The greatest Dutch engineering feats are not matters of clear choice but of necessity, at least to a great many citizens. In the quest for a healthy environment we legitimately expect many sacrifices from mankind, but to propose that the Dutch abandon their dams and dikes is like proposing they abandon their history, tradition, and a good part of the nation itself. Where the Dutch have had a choice, they have chosen not to pit their own energies against those of nature.

When the Dutch came to the New World in the seventeenth century with four hundred years of coastal engineering experience, they did not choose to confront the sea. Like the English, French, and Spanish settlers, they found not interfering with nature safer and, of course, cheaper. To Henrik Mey, Henrik Hudson, Verrazano, Cabot,

Raleigh, and other early explorers, most of the American coastline looked bleak and worthless. From Florida to New York they sailed up and down, restlessly looking for land more suitable for habitation than the low, narrow barrier islands. It was no accident that Raleigh settled his Roanoke Colony on Manteo Island behind the highest point of the Cape Hatteras barrier island where the forested dunes rose into genuine hills.

Unfortunately, he chose one of the few places where a permanent Indian population also lived close to the beach. For the most part, the Indians too had rejected the land near the beaches, and coastal islands in general, except for summer fishing camps. If our guess is correct, the demise of Raleigh's colony is one of the few examples of Indian-White hostilities in which the controversy is a piece of coastline.

On Cape Cod, where millions of tourists now visit the beaches and vie with homeowners for a view, the early colonists of Wellfleet, Plymouth, Barnstable, Eastham, Yarmouth, and Sandwich all chose sites near the bays and harbors. Farther south, the Dutch on Manhattan virtually ignored the beautiful sandy shores of Long Island to send their settlers up the Hudson and down into New Jersey.

For some two hundred years the colonists who owned beach property considered it a burden as often as a benefit. In 1690 Thomas Budd, a Quaker, gave away the present site of Atlantic City, New Jersey, for four cents an acre. While negotiating that transaction he talked up the price of inland property he was selling to ten times the price of the beach property.

This was at a time when America was an exploited colony, much too preoccupied with the labors of survival and agriculture to launch an aggressive attack on the most difficult of all environments to tame. Unlike the Dutch, the colonists were not crowded against a short coastline. But as we are well aware, the colony grew into a nation and into an industrial-military power more quickly than any nation on earth. Our conquests of wilderness and our human adversaries quickly bred the confidence necessary to face the sea itself. But it was not until Americans began to need an escape from themselves that they took to the beaches.

Unlike any other people in history, we flock to the beaches in search of play, relaxation, romance, and mystery. The population of Cape Cod grew 40 per cent between 1970 and 1976, while Massachusetts was only growing by 2 per cent. In Los Angeles on a hot day a million people seek refuge on some ninety public beaches. Most of Florida's beach development rests on narrow barrier islands or on land pumped up out of the bays and lagoons since 1950. As in every other nation, servicing such great needs has become immensely profitable. Although the recreation and resort industry can be said to be one of society's luxuries, it is sustained in its manipulation and development of the beaches by a popular following that is larger, more democratic, and more affluent than was ever enjoyed by developers and engineers of earlier times.

Like Holland's Zeeland, we will gear our response to the cost of coastal disasters. In 1977 almost $2 billion in relief funds were paid to victims of coastal storms. As we lose more and more of the Gross National Product on the coast, expensive and elaborate technology will seem more and more reasonable. When the major hurricane disaster comes, as it inevitably will, and rolls through the casinos of reborn Atlantic City or bulldozes ten thousand mobile homes on Florida's west coast, a billion dollars for seawalls, jetties, groins, bulkheads, and sand might seem both economical and humane. And of course such measures would bring jobs and boost the GNP.

This brings us to the midpoint of our book's argument. We have surveyed the natural system that creates and maintains the beaches and shown that the beaches are not mere sand dumps existing at the mercy and whim of a cruel sea. They are part of a large and complex geological band that begins three or four miles out to sea and ends high and dry next to the mainland proper. This entire beach system is married to the dynamic forces of the oceans and winds and melting glaciers. It survives only by unceasing motion.

How will we as Americans react to the danger? We possess both the most sophisticated technology on earth and the broadest scientific perspective on the natural system. How we use this skill and knowledge is a question whose answer lies in history and culture. Many observers have noted many aspects of American culture that resemble Rome more than Greece. Certainly our popular literature extols Dutch engineering a thousand times more frequently than their conservation practices. Our history as a technological society is much shorter than Holland's, but it offers several examples to prove that our cultural bias is toward armoring the coastline rather than working with the subtler natural processes.

9

Sand Castles and Supermen

THE FOUNDERS OF this country knew a good beach when they saw one. After defeating the British, Philadelphia's new elite began to sail their boats down Delaware Bay to Cape May, the southern tip of New Jersey. There, not far from a little Quaker village, they could anchor in the bay on the lea side of the cape and swim in the ocean. The beaches were broad, the sand firm, and the bottom gently sloping. These Philadelphia dignitaries and a humble postmaster unwittingly conceived the nation's first resort and the undisputed queen of beach cities for almost a century.

Ellis Hughes, the postmaster of Cape May, capitalized on the beauty of the beaches with this 1801 ad in the Philadelphia *Daily Aurora:*

> The subscriber has prepared himself for entertaining company who use sea bathing, and he is accommodated with extensive house-room, with fish, oysters, crabs and good liquors. Care will be taken of gentlemen's horses. Carriages may be driven along the margin of the ocean for miles, and the wheels will scarcely make an impression upon the sand. The slope of the shore is so regular that persons may wade a great distance. It is

> the most delightful spot that the citizens may retire to in the hot season.

The postmaster actually supplied accommodations somewhat less comfortable than his ad promised. Hughes's Atlantic Hotel was a barnlike one-room building. At night a sheet drawn across the room separated, we are told, the ladies from the gentlemen. During the day the braver tourists put on their old clothes, including old shoes, and waded into the surf. Swimming was not yet considered a proper sport, especially not for the "weaker" sex.

Ellis Hughes and his hotel stand at the beginning of American beach development, symbolic of the mixture of fantasy, prestige, and avarice that have ever since been the backbone of the resort industry. By 1830, six other entrepreneurs had opened boardinghouses at Cape May, and by 1850 they were part of the country's most famous resort. A speedy Philadelphia steamer regularly discharged loads of socialites, bankers, captains of industry, and Congressmen onto the piers. Presidents Pierce, Buchanan, Lincoln, Grant, and Harrison attended dinners and parties in the big hotels. No other beach resort in the world matched its prestige. Cape May was the place to see and be seen, if you could pay the bills.

Many of the ornate and graceful houses built in the city's prime still please the eye with intricate woodwork, airy porches, and towered roofs. The affluence and famous visitors are gone. Powerful politicians no longer come to Cape May to hobnob. In fact, the local politicians of Cape May go to Washington and Trenton to plead with minor bureaucrats for economic aid. Cape May, like all economic creations, was bound to be eclipsed one day by more modern attractions, but it might have stayed at least as attractive and viable as Nantucket, Martha's Vineyard, Key

West, or Southampton. Its decline seemed to run out to sea with its disappearing beaches.

Despite the usual shaping and reshaping, the beaches remained broad, firm, and popular through the nation's first century. In 1908, Henry Ford raced his newest-model cars on the gently sloping sands. Six hundred feet of dunes separated his roaring cars from the popular boardwalk. Nevertheless, like most beaches, those of Cape May were retreating in the face of rising sea level. The retreat of three to five feet a year was not unusual. It is the exact proportion expected from the rate of sea-level rise and the slope of the coastal plain in that region. With development behind the dunes, no one worried or measured. The town went on building itself into a small city that would one day have to be defended.

Through the nineteenth century a respect for the power of the sea permeated the town's development. Most hotels were built behind the dunes, and the steamboat pier was designed so that it could be taken down in the fall and reconstructed in the spring. Thus it was saved from winter storms and could be moved yearly as the shoreline shifted. The first boardwalks were merely boards laid on the sand and picked up before winter or a storm. The wisdom of such mobility was underlined by the two lighthouses that fell into the sea before 1859. A third was built a thousand feet inland and still stands.

Then in 1911, boat owners prevailed on the federal government to build jetties at the inlet to Cape May Harbor north of town. The jetties quickly stopped sand flowing south along the coast. The beaches at Cape May grew thinner and thinner. By the 1920s the town was building stone groins along its beachfront to reach out into the longshore currents and grab what sand remained. They had little effect. The beaches were now retreating twenty to thirty feet a year. The airfield was losing one of its runways. The lighthouse seemed doomed to follow the first two. The little town of South Cape May was washing away, taking with it two Roman Catholic convents and a radar station.

By the time the 1962 Ash Wednesday storm struck, the beaches offered so little protection that the ocean flooded the entire city. In response to that storm the U. S. Army Corps of Engineers constructed a massive rock seawall along one mile of the waterfront. But by 1968 Mayor Frank Garvey sent an emotional and urgent appeal to the Department of Housing and Urban Development. "Our community is nearly financially insolvent," the mayor said. "Our community has exceeded its allowable bonded indebtedness by some 2½ times. . . . The economic consequences of beach erosion are depriving all our people of much-needed municipal services." Cape May had also discovered a curious economic effect—the more protection you provide, the more you spend later. Inflation alone could hardly account for the fact that between 1963 and 1968 the city floated more beach-protection bonds than in the eighteen years before 1963. The town of 4,500 people were saddled with $241 of beach-protection debt for every man, woman, and child.

The mayor also noted that after only five years the seawall was crumbling and that a twelve-block section of town called Frog Hollow was living in a "pall of gloom." The thrust of the mayor's proposal was what he considered a brilliant engineering scheme. He had been watching surfers and noticed they rode the swifter currents near the groins. He reasoned that more sand would be moving there and that a fence or wall built perpendicular to the end of the groins would cause incoming waves to slosh over and leave their sand. The town had actually built one such sand trap out of tree trunks laced with cedar limbs. Two months and an eighteen-inch accumulation of sand proved it was Cape May's salvation, the mayor said. He asked HUD to put his city back on its feet by funding a series of his devices.

Like many an engineer's scheme, the mayor's sand traps had the fatal flaw of trying to stabilize the moving beaches. Neither HUD nor the Corps favored the sand traps. The town did get two more miles of seawall. In front of the seawall the Corps occasionally drops enough sand for short beaches. The waves, however, crash against the unyielding rocks and their full power is thrown back on the beach, which quickly yields and disappears. Meanwhile storm surges actually top the wall, and in 1974, with the seawall holding back a storm surge higher than the city streets, torrential rains were dammed up in town and turned Cape May into a great lake. The town suffered $2 million in damage and had to use sewage-plant pumps to clear the water. To add insult to injury, the sewage itself went out with the water, and the beaches had to be closed for an entire week during the peak tourist season. If the town continues to wall itself in, it must soon take the next step and raise the seawall. A model for this exists further north at Monmouth Beach. Here the wall is so high that many second-floor windows have no view of the ocean.

The most recent of the Corps' plans to revitalize Cape May calls for $24.5 million to pump up a new beach and build a sand trap on the inlet jetty. A constant pumping would move the trapped sand to the Cape's beaches. A once-thriving resort is now kept alive by the equivalent of forced feeding and a respirator.

In 1968, when Cape May's economic conditions and erosion problems both became critical, the mayor said, "We have finally reached a point where we no longer have beaches to erode." He predicted that "without a beach this community will die." The death of communities is no easier to define than human death. However, the decline and fall of the resort favored by the nation's founders and five Presidents fits nicely into the frame of the Bicentennial. The time scheme is mainly symbolic. The beaches, while always retreating, did not really begin to disappear until early in this century when the harbor jetties were built. From there, one protection led to another until the beaches disappeared and all that was left to defend was a dying town.

Cape May is the rule for beach development, not a nicely packaged exception to prove our thesis. Wherever affluent people could be lured to the beach, persuasive entrepreneurs built accommodations. With few exceptions the stories are the same. On the central New Jersey coast a romantic doctor and two friends repeated the Cape May story on a much grander scale.

In May of 1820, Dr. Jonathan Pitney guided his horse across the sands of Absecon Island—a complex of beaches, marshes, and tidal creeks where local salt-makers and oystermen scarcely saw anyone arrive by land and not many more by sea, though brigands and pirates had once found the area an ideal retreat from their own sort of hectic life. Dr. Pitney, a man from the hills and tight valleys to the north, immediately fell in love with the island and especially the spacious ocean vista. By 1845 he and a friend, General Enoch Doughty, stood on a hill overlooking the beach and the Atlantic. Although most inlanders saw the island as a "succession of barren sand hills and unproductive swamps," Pitney made the declaration that was both a blessing on the land and its doom. He said, "This should become the El Dorado of the Atlantic Coast, Enoch." The statement should go down in history as a model for dangerous leaps of imagination.

General Enoch Doughty and Dr. Pitney began their

beach development by recruiting railroad engineer Richard Osborne. Perhaps it is fitting that the country's first truly big resort city was launched by a romantic, a general and their hired engineer—and that this combination had no trouble attracting the millionaire backers it needed. Osborne immediately combined the investors' desires in a master plan. The finished drawing showed a new city fronting the ocean. Superimposed on the neatly drawn breaking waves were the gold letters ATLANTIC CITY.

By 1854, backers of Atlantic City had a going resort complete with bridge to the mainland, a railroad line to Camden, and a brand-new hotel. Atlantic City became an investor's dream where land that Dr. Pitney and his friends had bought for $17.50 an acre soared to thousands of dollars an acre. For those who could not invest personally, Parker Toys developed the game of Monopoly. (Monopoly cards represent properties in and around turn-of-the-century Atlantic City.) Visitors went to Atlantic City as much to see the place that had made men rich as to see the ocean. But the ocean and the beaches Dr. Pitney had loved were gone. The dunes had been leveled and replaced by the boardwalk and the seawalls. The marshes had been filled. The broad wild beach had shrunk. Today the city itself has almost unmanageable slums, and its life is sadly similar to that of the urban centers for which Atlantic City was to provide an alternative. That this El Dorado will rise phoenix-like and be gilded by the gambling casinos is unlikely. Development, whether casinos or boardinghouses, does not guarantee affluence, especially on the beach.

What hastened Atlantic City's decay were the engineer's plans—plans which were formulated to guarantee the investments of several generations of speculators. To protect the buildings from the ocean, the seawalls were necessary, and when the ocean began to break these apart a few years later, the U. S. Army Corps of Engineers replaced them with a sheet-steel seawall. Yet shortly after this, a mile-long jetty had to be built to stabilize the shore on the north edge of town, while the stone groins along the ocean beach were extended and new ones built. Both ocean and inlet beaches now require periodic pumping of sand fill to maintain enough beach for sunbathers to lie on.

While the nation's entrepreneurs were learning the lessons of New Jersey, a brief and gentle "boardinghouse era" occurred. In the last half of the nineteenth century, people who came first as boarders returned later to build stately white gingerbread houses, brown-shingled cottages, and grand hotels on whose spacious porches rocked the richest men and women in America. The great Oak Beach Hotel on Long Island boasted five hundred rooms. As early as 1830, vacationers were crossing one hundred miles of North Carolina's flat coastal plain and wide sounds to summer at Nags Head on Hatteras Island. By the time the Civil War broke out, that resort boasted a half-mile-long pier reaching into the sound to receive mainland visitors. The visitors had their choice of several large hotels, and when they wished to bathe, a mule-powered railroad carried them to the ocean beach. The developers of this first southern resort—following the example of the Indians, Sir Walter Raleigh, and a few small fishing villages—located on the forested shores rather than on the ocean beach or in the dunes. There they stayed until the 1930s, when the state built a highway along the oceanfront, giving official sanction to development that has been flooded and beaten apart by the sea several times.

The boardinghouse era had always been doomed by the fantasies nourished by Atlantic City, whose early success reduced a whole ecosystem to two images in the public's imagination: sun-warmed sand and gentle surf, and a Monopoly card—that prize in the pack, the Boardwalk. Simple ideas move fast, and they move most easily where the land is cheap and local resistance weak. The impetus for intense and sustained waterfront development moved like a wave from north to south. In the South the decay of the plantation system and then the Civil War left the warm southern beaches and sea islands ripe for exploitation.

Even before the Civil War, rich Northerners were visiting the Sea Islands of Georgia and South Carolina, buying financially embarrassed plantations for hunting preserves. When cold weather closed in up North, the sportsmen boarded trains and came South to shoot duck, deer, pigs, and alligators. The war ended this invasion, but only temporarily. As soon as the Union Army moved South, land speculators began their own war of conquest.

On South Carolina's now exclusive Hiltonhead Island speculators began intense wheeling and dealing two years before the end of the war. In 1863, shortly after the Union Army had marched in, Hiltonhead sales were declared open. Northern speculators bought over forty plantations at a dollar an acre, outbidding the blacks their army had come to liberate. Blacks, however, who had been working for wages and selling fish to the Union Army, managed to grab six plantations away from their white "liberators." Then, in 1865, General Sherman granted some five hundred thousand acres of southern coastal plantation property in South Carolina and Georgia to freed slaves, an order quickly modified by President Andrew Johnson in his efforts to unify the nation by appeasing southern landowners. Johnson granted to Confederates swearing loyalty to the Union the right to return to their lands so long as they did not displace any newly established black landowners. When the Redemption Act allowed white landowners to buy back their lands for back taxes plus interest, many could not raise the money or they discovered that the deeds proving ownership had been burned in the courthouse. Into this situation moved F. R. Klem, Roy Rainey, and W. L. Hurley, Yankee land speculators. By 1936, almost all of Hiltonhead Island was in the hands of Yankees, the original population of three thousand blacks reduced to three hundred. The land was now in hands willing and able to transform it as they pleased.

Though the maritime climate and its safe bayside harbors made Hiltonhead desirable, few nineteenth-century speculators were interested in the oceanfront. Symbolic of the natural hazards that warned them away was the great hurricane of 1893, with a surge that topped most of the island. So sudden and total was the devastation that Clara Barton, founder of the Red Cross, came to guide relief efforts, and Joel Chandler Harris, who recorded the Sea Islands' Uncle Remus stories, memorialized the storm's destruction in the pages of *Scribner's Magazine*. The passage of that storm became a date from which islanders measured marriages, births, and deaths—so many years before or after the "Big Blow."

The southward-moving wave of beach development only reached Florida after the turn of the century. Those were the years during which the internal combustion engine and the automobile placed incredible power in the hands of almost every ordinary American—if not in his hands, at least in his imagination. Individual men were now able to speed through the landscape and even through the sky. And of course, the powerful little engines were bolted into speedboats, which began to ply the broad sounds and bays of the East Coast. The sudden success of these powers meant that no environment could forbid people entry or resist their inventive efforts at domestication.

In this state of mind twentieth-century men and women

found a few nineteenth-century people struggling with the mangrove forests, sultry swamps, and insect hordes of a barrier island which protected the little town of Miami from the ocean. Miami Beach is one of the few major American cities to be born full blown out of the twentieth century, a town without a history or at least a heritage. It doesn't even have much folklore, though its real-life characters are often as fantastical as any in legend.

The first nineteenth-century "developer" visited in 1870, back when agriculture was still king of the countryside. He saw an island where a sandy ocean beach with low dunes shimmered in a sun that shone year round. Behind the dunes an impenetrable, semitropical forest testified to the fertility of the land. To a man who knew good soil, Miami Beach appeared to be a superfarm yielding untold riches. And right he was too, although Henry Lum, born a Pennsylvania farm boy, was not looking at the superfarm we now know existed in mangrove swamps and marshes. He imagined an island cleaned of its jungles and transformed into an orderly plantation of row crops.

Henry Lum was ready for any challenge. All over the country he had seen the wild lands transformed into farms. He himself lacked none of the imagination that for better or worse drives people to try the impossible. He had spent his life trying to make his fortune in the American way of hard work and speculation—in various combinations of each. He had gone West with the Gold Rush but had found more money selling bread to the miners than hacking away at the hills. Then he had returned East and begun tree farming. By the time he and his fifteen-year-old son Charles sailed to Miami Beach in 1883, he had twice made a going business (bakery and tree nursery) out of nothing. When he saw three isolated coconut trees at the south end of Miami Beach, they seemed next to nothing too, but they whetted Lum's imagination. He attracted a number of northern investors and planted some three hundred thousand more coconut trees. After his death several years later, his son Charles and his wife tried to pay off an eighty-dollar debt to a mainland innkeeper with ten acres of land. They were laughed out of the inn. The prospects out there by the ocean seemed dim for everyone. Though the plantation failed, it attracted other investors who staked out an even larger plantation up the beach. They too failed.

The island the coconut planters had tried to get rich on was the last barrier island in the southernmost string of East Coast barrier islands with sandy beaches. Behind the beaches and the dunes were mangrove swamps, sometimes a mile wide. They were fantastical tangles of roots, debris, and mud, but also unimaginably rich breeding grounds for fish and shellfish—as well as for the clouds of minute black flies which helped doom the coconut planters by driving away their laborers. If the island was to amount to anything in human terms, it would have to be conquered, and the kind of investor who had been attracted to the get-rich farm schemes was exactly the kind of person who had conquest in mind.

Charles Lum gave up his father's plantation while down the beach two other wealthy planters, Ezra Osborn and Elnathan Field, realizing no profit, abandoned an even larger spread. As the irony of history would have it, John Collins, a horticulturist who had bought in on the Field and Osborn farm was the first to see that the beach that could not grow coconuts could yield other kinds of profit. Where plants did not thrive, perhaps more adaptive human beings would.

Collins was prompted to see the potential for Miami Beach by his family. Alarmed at the father's fruitless investments, first in coconuts and later in avocado farming, his children, the Pancoasts and Collinses of New Jersey, packed up and came South to supervise their seventy-two-year-old patriarch's horticultural investments. They envisioned their father's island being made into a boardinghouse resort like Ocean City or Seaside Park, New Jersey. Their vision may have been shaped by two facts of their northern heritage: first, they were wealthy; second, they were Quakers. Because they were wealthy, they had been in the habit of passing their summers on the highly developed Jersey coast. As Quakers they preferred quiet, residential life to busy hotels and casinos. They brought with them the nineteenth-century's boardinghouse vision. But wealthy as Collins and his children were, and modest as were their dreams, they could not defeat the beach alone.

Their great wooden bridge across two and a half miles of Biscayne Bay was only half finished when they ran out of money. In November 1912, the world's longest wooden vehicle bridge—abandoned—stood as a symbol of the costs and complexities of developing the coast, but also of the enormous will to succeed. In those early years of this century, as in classical Rome, no costs seemed too great and no natural threat too terrible to stifle mankinds' desires for long.

The Collinses and Pancoasts were rescued by a man of truly Roman appetites—a young Indiana millionaire named Carl Fisher, who had already founded the Prest-O-Lite Corporation and the Indianapolis Speedway. By 1912, not only had Fisher proven himself as one of the great mechanic-farmboys of the automobile age, but also he had demonstrated his complete faith in the power of technology. Fisher had come to Miami to race his new speedboats in Biscayne Bay. The warm weather and quiet waters of the bay suited him fine, but he was annoyed by the garbage that kept drifting around his pier. To protect himself from the litter of Miami's growing population and increasing number of tourists, Fisher hired a dredge to build up an artificial headland which would divert the garbage. No sooner was the headland built than Fisher realized what he had done—he had created land. As a real estate investor he immediately recognized the profit in creating land rather than buying it. His Point View Subdivision, not far from his pier, soon rose out of the bay to become the first step in the filling and polluting of Florida's rich estuaries. Fisher proved that land production could be carried out as methodically as car production.

The scale of Fisher's faith in technology at that time is measured most clearly by his proposal to build the first interstate highway, the "Coast-to-Coast Rock Highway," a project he proposed to finish in three years. The loan of $50,000 to a stalled bridge-builder whom Fisher called "a bantam rooster, cocky and unafraid," was a trifle compared to the multimillion-dollar highway and most of Fisher's investments. The bridge itself was little more challenging than Erector Set engineering. The island at the end of the bridge drew Fisher like the end of a rainbow. He could no more resist the temptation to transform the entire island than he could resist the lure of beautiful women, fast cars, and speedboats. Even before the bridge was completed, Fisher's speedboat had roared into Collins' sand-banked canal on the marshy side of the island and he envisioned a city of pleasure domes rising by the ocean. (During this first trip his boat's wake collapsed the sides of the canal—framing Fisher's vision in one of the engineering mistakes that would grow apace with the city itself.)

Fisher's plans called for almost complete redesign of the landscape—and in fact he billed his "tropical garden" as a new Atlantic City. In 1913 the idea of remaking a mangrove forest and pumping hundreds of acres of dry land out of the bay awed a public that had not yet comprehended the power of modern technology. Fisher never had doubts. In short order the state of Florida and the U. S. Army Corps of Engineers granted him permits to do what he would with the swamps, the marshes, and the bottom of Biscayne Bay. When the palmetto and mangrove roots stopped his laborers, he had a special machete plow built and shipped down from the steel mills of Indiana.

Will Rogers says that Fisher made the dredge the "national emblem of Florida," and that Fisher's only test of the new land was whether the sand would hold up a real estate sign. At first, some of the new land wouldn't hold the signs, and in fact two mules drowned in something akin to quicksand. Twice they were buried, as law required of dead mules, and twice (with a horrendous smell) they floated up again. They were finally carried to the mainland, where they still rest in peace. The development went ahead.

Even before the first disaster there were prophetic warnings. By 1926, Biscayne Bay had been spanned by bridges in several places, and a number of artificial islands had been built to support the bridges, houses, and boat slips. That year Commodore Ralph Monroe, who like other locals recognized that the beauty of the Florida coast was born of the ingenious engineering of wind and tide, temporarily saved the bay from further filling. Monroe, one of Florida's first conservation activists, obtained a court injunction against further creation of artificial islands, claiming that the islands, causeways, and jetties would dam up hurricane tides and cause disastrous flooding. His warning almost immediately became fact.

On September 17, 1926, a hurricane challenged the technology of Fisher's "tropical garden city." By midnight the storm had driven a powerful surf over most of the island and through the doors and windows of hotels, houses, and real estate offices. A bizarre debris of oriental rugs and fine furniture washed toward the mainland. The next morning torrential rains in the Everglades compounded the tidal flooding. Water could not escape the bay through its old inlets, channels, and washover points. Not only Miami Beach went under, but also the new subdivisions along the mainland side of the bay and many parts of Miami itself that had never before been flooded. The very first stages of development of Miami Beach must share a good portion of the blame for the 854 Miamians hospitalized in that storm and the 25,000 left homeless.

Carl Fisher, hurrying down from his new development on Long Island's Montauk Point, found that on the Beach itself his lavish hotels had suffered almost no structural damage. Fisher wrote to his former wife:

> Dear Jane,
>
> Don't worry about the hurricane. It will prove to be the greatest blessing to Miami Beach in the long run. Miami Beach will be more beautiful, bigger and better than ever.

Fisher's predictions of *bigger* came true—although the debate over "more beautiful" and "better" remains one that frequently comes to life in city council meetings.

The means of making the beach bigger and better was to fortify it. Carl Fisher had started the process in 1918 when a northeaster tore the sand from under one of his oceanside Roman pools. At that time he had built Miami Beach's first groin. After the hurricane the city council moved to protect beachfront homes by extending Fisher's groin system and declaring a bulkhead line at the high-tide mark. This government pattern of aiding the rich while ignoring the responsibility to help those who had suffered most—the residents on the Miami mainland—would later become standard in beach disasters. Meanwhile construction was the beginning of the end of the beach at Miami Beach and elsewhere.

The story of Miami Beach now has become the story of scores of cities along the Florida coast. Everywhere the bays were deepened, the beaches bulkheaded and seawalled. Groins and jetties began to poke out like a million pins in an elaborate hairdo. All over Florida beach land prices soared 25 to 100 per cent a year.

Like Fisher, the people with money have been the people with technology and the greatest interest in hiring technology to stabilize their property. Nor have they been content to merely maintain the status quo. After World War II when the real estate boom sprang from the embers of depression, entrepreneurs pushed technology not only to stabilize the shoreline but to extend it seaward. They convinced the Miami city council to pass patently illegal legislation extending the bulkhead line seventy-five feet seaward onto the tidal land owned by the state. Citizens, seeing in this a giveaway of their beaches, took the council to court. But six years later when they had finally won their case, the great postwar luxury hotels, including the Cadillac, Sans Souci, and Saxony, had already extended their domain across public land. They were not forced to retreat.

The court victory was one of principle only. Bulkheads and groins were still allowed up to the high-tide line, and ever since the citizens won their case against extending these structures seaward, the town council has allowed more and more bulkhead permits—applications often accompanied by cries of distress and imprecations against the foul work of nature that pulls the beaches out from under expensive motels, hotels, and pool patios. But at least the groins accompanying the bulkheads have stopped. Yet the longshore transport of sand has also been stopped, and ocean waves must concentrate all their energy on what's left of the beach in front of the bulkheads. Since the mid-fifties one of the great surprises for visitors to Miami Beach has been the absence of beach, especially at high tide. Even at low tide, steel and timber walls built out into the longshore currents to capture sand also kept many strollers away from in front of the luxury hotels.

In 1967, people who could not see the beach through the hotels nor climb over all the groins to walk the beach, raised so much ruckus that the U. S. Army Corps of Engineers offered a plan to build a new beach in front of the hotels (at taxpayer's expense) to replace what had disappeared due to the actions of the resort developers. The Corps proposed importing enough sand to create a beach seven miles long with an eleven-foot-high dune in front of the hotels. The shoreline was to be some three hundred feet seaward from the old bulkheads. The $35 million construction cost did not include annual sand replacement. But if the Corps' proposal was at first unacceptable to taxpayers because of the high cost, it was also unacceptable to the hotel owners for another reason.

The Corps insisted that any and all taxpayers must have access to a restored beach even if access is through hotel lobbies. Almost as soon as the county allocated funds for the project, hotel owners went to court to seek damages, claiming the restoration of public beach would hurt business. While the plan was debated the ocean continued to edge closer to the hotels, threatening to undermine pools, patios, and expensive rooms with ocean views. Miami Beach began to lose tourists. Within a few years most opponents of the new beach began to think new sand might be a good idea. Hal Cohen, director of the local Tourist Development Authority, promoted the new beach as a way to "change the whole image of Miami Beach" and save it from Atlantic City's problems.

By May of 1977, Great Lakes Dredge and Dry Dock

Company, under the directions of the Corps, had anchored its giant orange barge off Miami Beach. Sucking sand from the ocean floor, it spewed it out again in front of the hotel bulkheads. By 1980 the Corps hopes to have built a beach 300 feet wide and 10 miles long for $64 million. Almost 14 million cubic yards of sand will have been moved from the continential shelf onto the new beach, making this the largest beach-building project in the world. Once built, it will cost $1 million a year to maintain. It will also lose 211,000 cubic yards of its new sand each year, and while the Corps will not block hotel views with a re-established dune system, they are dumping an extra measure of sand at the high-tide line.

While Miami Beach business people looked toward a brighter future, in the murky depths just offshore their undoing may be the very hole from which the dredge sucked their new beach. Even though the dredge is mining sand as far as two miles offshore, such changes in the profile of the inner shelf will affect the beaches. Dr. Victor Goldsmith, of the Virginia Institute of Marine Sciences, feels that the changes will depend on the shape of the mined area and the direction of the waves passing over it. In other words, the beach may show no effect for a while, then with relative rapidity begin to make its adjustments.

Oceanographer Conrad Neumann, of the University of North Carolina, points out what may be a more fatal flaw. The new sand, he says, is mostly made of the skeletons, shells, and excretions of small marine organisms that live and die in the quiet depths. Their fragile remains will not withstand pounding surf as a quartz sand does.

Miami Beach's original sand contained much more quartz carried southward by thousands of years of longshore currents. Man-made beach erosion has left most of this sand in a shoal of Key Biscayne's Turkey Point. This same sand is being sold to Miami Beach developers to build hotels and condominiums and probably to strengthen the seawalls which will hasten the erosion of the new beach. Dr. Neumann calls this whole process "uncycling."

Despite its short history and many problems, Miami Beach became the model for commercial development on both the East and West coasts. Along almost every open beach south of Boston on the Atlantic and south of San Francisco on the Pacific, high-rise buildings and motels rise in the teeth of the ocean. Since beach property now costs thousands of dollars per front foot, the more people you get on each foot the more money you stand to make. In many places condominiums, hotels, motels, and apartments are rising like an impermeable seawall. A recent novel in which a high-rise condominium collapses during a hurricane is sensational and sometimes misinformed; but even so, it fails to convey the real enormity of the disasters waiting to happen.

We know only two solutions for places like Miami Beach: more technology or less development. Technology seems to provide only short-term answers. Much of it provides protection from the angrier moments of the sea only at the expense of sacrificing the sandy beach itself. Less development, or none, seems too drastic to consider. It is quite likely in these areas that politics and short-term economics will mandate protection. That protection will be designed to guard development against storms. Eventually it is all subject to the rule of sea level. As the sea continues to rise, the protective walls can only get thicker and higher and more expensive. The only tourist activity will be strolling on top of the seawalls to see the ocean.

On less-developed sections of the coast local governments and property owners often reassure themselves of their security with a kind of golden mean—perhaps more appropriately called the magic setback. Set your development behind the first dunes, they say, and you'll be snug and safe. Coastal residents were hardly snug when in February, 1973, a relatively small storm moved down the Atlantic Coast to Cape Hatteras, North Carolina. Residents saw twenty- to thirty-foot waves carving into the beach and dunes like the sharp blades of an army of bulldozers. During three tides the surf swarmed over the dunes. Motels and cottages collapsed. Those built on the high "stabilized" dunes collapsed when the waves, unable to wash over the dunes, vented their full fury on the dune face, carved away the sand and undermined everything above. Buildings constructed on pilings, the "safe way," collapsed as the waves dug the sand away down to the base of the pilings. In other places, so much sand was pushed under the buildings that waves could no longer pass beneath the pilings and the houses were flooded or floated away. Many buildings were never found again.

The United States has not yet committed us to a Dutch-style, totally engineered shoreline, but we are moving that way. Congress, seeing a tremendous pork barrel in the always popular beach projects, took the first step in 1972 when it passed the Rivers and Harbors Act, which called for a policy of technological experimentation for beach protection. Fortunately, the President killed this measure when he pocket-vetoed the entire bill.

The U. S. Army Corps of Engineers now finds itself caught between its old engineering bias and a new emphasis on ecology. Congress, despite the Corps' desired direction, will ultimately determine what projects will be undertaken, and Congress likes to do things for its constituents that are immediately visible and that yield votes every two years.

On what will all the big earth-moving machinery and construction crews be used, now that interstate-highway building is slowing and the construction of the great inland dams and waterways is being defeated by environmental concerns and rising costs? Machines which have leveled mountains can also push back the sea.

All over the country big money is urging the public to put its faith in engineering solutions to coastal problems. Private investors own over 85 per cent of the Atlantic coastline and over 60 per cent of the Pacific coastline. The demand for beach property continues to soar because, of all recreational land, the narrow strip of beach is rarest and most desired. All coastal states feel pressure, therefore, from the same investors who persuaded the Miami Beach zoning board to grant so many questionable variances in 1967 and 1968 that the state attorney general declared 95 per cent of them illegal. Controls over beach development which make politicians and not environmental realities the final arbiters, will always fall prey to big money, and big money almost always prefers big technological solutions. Development itself is technology, and the history of seacoasts from Carthage to Miami makes it clear that it takes technology to protect technology.

We are not simply talking of bad investments. As with the dredging and filling of Biscayne Bay and the wall of hotels on Miami Beach, the public pays the cost—in disaster aid, shore protection, and devastated natural resources. Worse yet, the developers, some honestly ignorant, some willfully ignorant, put many innocent American in economic straits and physical danger.

Hundreds of thousands of Americans live in a situation similar in many ways to that of the Netherlands. Sometime a disaster worse than the 1900 Galveston hurricane or the 1953 North Sea blow will hit Miami Beach, the west coast of Florida, the south shore of Long Island, the barrier islands of North Carolina, or the Gulf beaches of Mississippi, New Orleans, or Texas. The inhabitants can only be protected by technological means. To say, "Let them perish," would require a harder heart than most Americans have.

America, with all the arrogance and power of a modern Roman empire and with a sentimental attachment to the sea, stands on the verge of a Dutch-style commitment to coastal management. To avoid that commitment means to try to change the thrust of history and the American mind. We are at the end of a long cycle. We began as pioneers eschewing the beach in the way of primitive people; we became developers; developers in trouble turned to science and engineering, and because of that alliance, development has boomed. We ought to rethink our development, and perhaps common sense behind earlier peoples' humble respect for nature might benefit us today.

This fork in the road has inspired many engineers, as well as homeowners, motel operators, and mayors, to invent new devices and processes for protecting the beaches. The inventors and those who hope to benefit from their work say the new devices imitate nature. Many inventions, like the beach itself, are temporary, adjustable, and movable. They acknowledge, in their attempt at imitation, that the natural system works. The new wave of shoreline engineering seems to operate on the old maxim: "If you can't beat 'em, join 'em."

Engineering is not really ignorant meddling. Shoreline engineering is a complicated exercise in physics and construction practiced by many exceptionally intelligent technologists. Most engineers attempting to protect beach development are well aware of the problems caused by the traditional jetty, seawall, or bulkhead. They understand the mechanics of waves and longshore currents. They are far ahead of the public imagination in using this knowledge and the materials of the space age to create a new kind of engineering, one which they feel is more natural.

There are many man-made solutions to coastal problems. We will look next at both the tradition and what appears to be a new trend toward imitating nature. The question is, Are we on the verge of a new technology that will in fact give us the powers of supermen, and can we tame a sea creeping always closer to development that is no more resistant than sand castles?

10

Engineered Shorelines: The Point of No Return

THE MOST AMBITIOUS proposal for establishing a status quo on the U.S. coastline originated in Singapore. In 1971 the Housing Board of that incredible human beehive had to pump new land out of the sea. The board armored the reclaimed stretch, 5 miles long and up to 2,200-feet wide, with a seawall. When waves began to undermine the wall, Siew-Koon Ho, a board member and engineer, proposed a more natural protection—artificial headlands. If the rocky headlands of a natural coast drew wave energy and protected quiet pocket beaches between them, engineers could create similar structures, Ho reasoned. Singapore's headlands of rubble reach out into the longshore current and trap sand. The result, at least temporarily, is a scalloped shoreline, a series of low, triangular headlands 800 feet apart with quiet bays between. In 1974, Ho and an Australian colleague, Richard Sylvester, described the project in *Civil Engineering*, the magazine of the American Society of Civil Engineers. Sylvester said that here, finally, was the way to beat the Atlantic Ocean on North Carolina's Outer Banks.

For the Atlantic Coast Sylvester and Ho propose growing the headlands behind floating breakwaters. A breakwater is simply a wall parallel to the shore. Behind it a wave shadow forms and sand drops out of the weakened longshore current. Sylvester and Ho propose an ingenious movable breakwater built as a platform of submerged steel tubes. As sand accumulates behind the breakwater, the platform is moved seaward until the headland reaches the desired size. Now the headland is stabilized with a rock-armored tip and flanks.

These two engineers, internationally respected by their colleagues, point out that a singular advantage of the protective headlands is that the dunes along the beach can be bulldozed and the sand used to fill in behind the headlands. The newly leveled dune area is planted with trees and lawn grasses which would now thrive, their roots finding the water table not far below ground. Both the old dune field and the new beach become prime development and recreation areas and broaden the local tax base as the shoreline becomes stable.

The headland system also offers beach users the advantages of many types of beach within a short distance. Near the headland itself wave energy will be strong and the waves right for surfers. In the cup of the bay waves will be smaller. This is an area for weaker swimmers and a calm place where people can launch boats. A beach with a series of bays would allow fishermen, swimmers, and boaters to enjoy special preserves.

Sylvester and Ho like to point out that all these benefits are the result of engineering that imitates nature. They say their system grew out of observations of beaches adjacent to natural headlands. Their vision of engineering which imitated and improves on nature has spread fast among their colleagues. Orville T. Magoon, Chief Coastal Engineer of the South Pacific Divison of the U. S. Army Corps of Engineers, calls artificial headlands "one of the most promising means of stabilizing a shoreline and also developing a recreational beach." There is no reason, he believes, why engineers cannot build structures that "will duplicate the success of nature in stabilizing sedimentary coastlines."

Engineers have often tried and failed to protect development by stabilizing the beach. History tells us that betting on beach stabilization is a high-risk wager. Like pulling the handle of a one-armed bandit, it's hard to quit betting, especially if big losses are preceded by small wins. But we don't mean to cast aside either science or engineering with this metaphor of slot machines any more than we would suggest ignoring your doctor because George Washington's doctor used leeches. Nevertheless, we know that no science is born without an intellectual and emotional heritage. The important question for us to ask about modern coastal engineering is not whether its inventions are different, but whether they spring from the same perspective and bias that we have traced from Rome to Holland and across the Atlantic to the United States.

Engineers of the past twenty years, many of whom are still teaching and practicing, have the highest failure rate of any group of engineers in the world. If bridges were failing as frequently as seawalls, a drive on any major highway would require considerable disregard for life and limb.

Most shoreline engineers would dispute these odds and name a number of places where they have tamed the shoreline. Lim Vallianos, an outstanding engineer with the U. S. Army Corps of Engineers, has argued with us that we have overlooked hundreds of successful shoreline projects all over the coast. But a shoreline engineer considers a jetty or seawall a success if it endures fifteen, twenty, or thirty years, its predicted design life. Also, to Vallianos and other engineers, a seawall that lasts fifteen years and protects the property behind it is a success whether or not it destroys the beach in front of it. Cape May's jetties successfully protect the harbor inlet, although the town spent itself into poverty trying futilely to hold its beaches and tourists, and to defend its property tax base against the ocean.

Developers and engineers frequently talk about environmental "trade-offs," giving up some natural quality to gain another benefit. The trade-offs required by many coastal projects, however, are seldom factored into the engineers' cost analysis or even predicted. Engineers, trying to protect development, volunteer for what is usually a no-win situation. The beach changes in shape and position because it is part of a dynamic equilibrium. Though this phenomenon has not been canonized as a law of nature, it is an indisputable fact of life and geology. Engineers are trying to stop the beach from responding to the dynamic equilibrium, to remove materials and shape from the four-way series of trade-offs nature requires within her own system.

The following principles have made engineers indignant and angry, but they are proven by careful observation of coastal engineering and its consequences:

1. No erosion problem exists until people lay out property lines and build. Beach changes only trouble people who have strong attachments to immovable objects and fixed lines. Shoreline engineers are rarely, if ever, called in to stabilize a wild beach.
2. Anything built on or near the beach usually increases the rate of erosion. Seawalls, bulkheads, groins, and house foundations reduce the flexibility of the system to respond to changes in the dynamic equilibrium. If energy patterns and sea level change and the beach does not, residents will lose more beach than under the natural system.

3. Once you start protecting the beach you can't stop. By destroying the beach, most protective measures eventually create peril for themselves and increased danger to the development they protect. When protective seawalls or artificial dune walls or jetties begin to give way, a healthy natural system does not exist as a backup. Also, residents seeing a seawall crumble often feel more frightened of the ocean than those watching natural erosion. The sequence of beach protection is also a sequence of increasing expenses.
4. In order to "save" the beach, we destroy it. When we stop erosion it is not to save the beach, but the development behind the beach. The beach that has become inflexible disappears rapidly, as it did at Miami Beach.
5. The cost of saving property is greater than the value of the property saved. This is particularly true when the long-range costs are accounted for and we recognize that storms occasionally wipe out whatever defenses have been erected.

Shoreline engineering is brought into the natural system by the people who are responsible for creating the problems, and their solutions usually cost taxpayers more money than the property behind the shoreline is worth, especially since the beach is often destroyed by its fortification.

Over the past three thousand years shoreline engineers have worked toward the same goal—stabilization. With many variations in design and theory, they have tried to realize their goal in one of three ways: building structures perpendicular to the shoreline (groins and jetties) to trap the sand in the longshore currents; building structures parallel to the shoreline designed to dampen wave energy before it reaches shore (breakwaters) or to absorb the impact of the breaking wave (seawalls, bulkheads, and revetments); replenishing or "nourishing" the beaches by pumping new sand onto the beach or into the longshore currents.

Groins and Jetties

It seems simple enough. Early peoples probably observed that wherever a boulder or log lay on the tidal land more sand accumulated on one side. On the side with less sand often a little hollow held a pool where snails or small minnows waited for the returning sea. If you build a wall of rocks, or of lumber, concrete, steel, or sandbags out into the surf perpendicular to shore, you have built a groin. (The name does not derive from any pelvic similarity but from the old French word *groyne*, which means "snout.") On the upstream side of the wall the sand builds out toward the seaward end, enlarging the beach. Chances are this sand is needed on down the coast.

If you or your neighbor want to save the next section of beach from sand starvation, another groin is needed. So on down the line. After the first groin was put up at Miami Beach, where sand moves from north to south, shoreline owners to the south began to put up their own walls, until the beach itself looked like a series of backyard fences. After groins of big sand-filled plastic bags became popular, sunbathers began to sprawl on them. After all, until the recent sand pumping began, the beach had disappeared in front of most commercial establishments.

The northern New Jersey shore, which has been armored much longer than Miami Beach, displays several generations of groins. When the tide is low, rusted metal walls and broken posts testify that the ultimate fate of groins is to be torn apart by storm waves. The debris and remnants are interesting as a kind of beach archeology, but they are, for all practical purposes, a form of garbage embedded in the beach and dangerous to swimmers.

Recently engineers have tried to preserve the life of groins and imitate the natural system by creating permeable structures. These may be latticework walls or weirs, giant jacks made of reinforced concrete or perforated steel. In some situations the groins are movable, so they can catch sand moving in one direction in summer and be taken apart when winter reverses the direction of dominant longshore currents. Sandbag groins are easily filled with a sand pump and easily moved. They are also easily destroyed by kids with a penknife.

The problem remains the same under all conditions—the groin is trapping sand that is moving somewhere. Wherever longshore currents move sand, the natural beaches have depended on this means of supply. Interruptions that build one section of beach erode another.

As with any device built and operated so simply, groins have been used by many people who see nature simplistically and for whom time is an unimportant dimension. In fact, in many parts of the United States groins were once more successful than they are today. The U. S. Army Corps of Engineers points out that much of our former sand supplies that found their way into the longshore currents to be trapped later by groins are now locked up in shoreline development, retaining walls, and inland dams—many built by the Corps itself.

Along California's Newport Beach the groin system can no longer wrestle enough sand from the Pacific to cope with seasonal changes in wave direction—summer waves from the south and winter waves from the north. Eddy basins and flood-debris traps on the region's rivers are collecting the coast's sand, and the submarine canyons swallow much of what does arrive.

It would seem logical to argue that once sand "fills" the groin by building out to its tip, the longshore process is restored by spillover. There is some truth in this, but the dynamic systems allow no such simplicity. The groins and the wedges of sand building up against them change the shoreline's shape and wave-refraction patterns, which in turn change pattern of longshore currents. In any event groins force the river of sand into deeper water where much sand can be lost, so that net sand transport may be much lower even after it begins to bypass a filled groin. Since the 1940s, along Miami Beach where the sand supply is scant, the beach was never much more than little pockets of sand and debris nestled in the corners where the groins meet the seawalls.

Small as they are, groins can affect beaches for miles along the coast. Their much larger cousins, jetties, have been responsible for the disappearance of whole towns. A jetty is built at the entrance to a harbor or at the mouth of an inlet to prevent sand from filling the channel. Under most conditions the currents of an inlet interrupt the longshore drift and the ocean side of a harbor channel is flanked by shoals. Sand does flow across the channel, however, by passing as a submerged crescent-shaped bar. The journals of explorers frequently record sending out small boats to locate this bar and sound its depth before the mother ship crosses into sheltered bays. A jetty tries to stop sand from building up across the mouth of the channel, or at least to move the bar to deeper water.

The length of a jetty, sometimes up to two miles, is determined by the depth of the channel it protects. It must guide the groove of the channel until it plays out on the deeper shelf. Engineers often sell jetties to local people as the solution to harbor problems by creating a stable channel, reducing dredging and offering safe entry. The situation is seldom so simple. Sand still forms the sides of the channel and many inlets continue to move, even within their artificial walls. Barnegat Inlet in New Jersey is used by almost half of the state's forty thousand small-boat owners and charter boats, but since being protected by jetties in 1940, the inlet has been a subject of controversy. The channel continued to shift and shoal inside the jetties, causing numerous boats to run aground on the jetties themselves. Users have now convinced the U. S. Army Corps of Engineers to do extensive dredging. At Wrightsville Beach, North Carolina, the Corps built a jetty to protect Masonboro Inlet at the south end of the island. Shortly after construction the channel migrated toward the jetty and larger boats had to hug the boulder wall to reach the harbor.

Even where a channel is safely maintained for larger boats, a jetty may disturb wave patterns and currents so much that small boats approaching or leaving the channel run into a local field of crosscurrents and dangerously choppy seas.

Whatever the problems inside the jetties, they do stop the sand. The story of Cape May, New Jersey, where jetties drove the town to poverty, is repeated in some form almost everywhere jetties have been built on the ocean beaches.

The jetties at Miami Harbor have been starving Key Biscayne, once home of the Nixon Summer White House. The Corps labels erosion "critical" all over the island. The continental shelf in front of the key is growing steeper, and the beaches lose some 100,000 cubic yards of sand each year. Because so much sand flow from the north is interrupted first by Miami Beach groins, then by the Miami Harbor jetties, the Cape Florida Lighthouse at the southern end of the key would have collapsed in 1969 had not the Corps placed a 283-foot revetment in front of it. Some little sand which escapes the jetties and groins forms a shoal in front of the lighthouse. This is that body of quartz sand that we noted could have been vital to offset erosion in the Miami Beach area, where the Corps is now using offshore sand that is almost 90 per cent animal skeleton—too fine and light to resist wave action. Dade County, however, sold the quartz sand to the Deroucher Construction Company to mix with the concrete of yet more development. The island's population doubled from forty-six hundred in 1970 to over nine thousand in 1977.

In Texas, where longshore currents are vital to the dynamic equilibrium of the beaches, jetties have a drastic effect. Robert Morton of the Texas Geological Survey recently completed a study of over 100 years of shoreline changes and has concluded that Texas jetties have trapped half the sand that could be in longshore currents.

On the West Coast, where the continental shelf has always dropped off steeply, where submarine canyons approach close to shore, and where the shelf itself supplies little sand to the beaches, any interruption in longshore currents can create rapid and radical change.

At Port Hueneme, a short distance north of Santa Monica, the U. S. Navy literally dumped the area's beaches down the Hueneme Submarine Canyon when in 1938 it built two short jetties to protect the channel leading to a U. S. Navy harbor. Before the jetties, considerable sand, swept along by strong, longshore currents moving eastward, had managed to cross the canyon. Once the jetties were built, sand was forced out into deeper water where wave-refraction patterns weakened longshore currents. The sand that should have supplied beaches further along the coast dropped into the nearby canyon instead. Then, to protect the narrowing beaches from further erosion, a 3,000-foot seawall was built east of the jetties. But this only aggravated the problem. By 1948, the beach behind the seawall, and even part of the adjacent dune belt, had eroded some 700 feet.

The U. S. Army Corps of Engineers still lists this as an area of critical erosion and recommends more groins and beach fill at an initial cost of almost $2 million per mile.

In Oregon, where the sea is even more powerful, a beach community once planned as the West Coast's most luxurious resort was entirely destroyed by a jetty built at the mouth of Tillamook Inlet. The site of the resort was a four-mile-long spit running north from Cape Meares and sheltering behind it the shallow and fertile waters of Tillamook Bay. The spruce-covered hills of the spit rise a height of 140 feet and once seemed at least as solid as the lowlands around the mainland side of the bay. Certainly the spit offered much greater tranquillity and more beautiful views than the lowlands. In 1907, T. B. Potter Realty, of San Francisco and Kansas City, announced plans for what Portland's *Oregonian* called:

> This most attractive vacation spot, eclipsing all the resorts for tourist travel in Europe, excelling anything on the Atlantic Coast and outstripping the best on the Pacific shores, is to be only a bit over a couple of hours' ride from the very heart of this city [Portland]. E. E. Lytle, who is building the Pacific Railway and Navigation Company's line, states that he is under obilgation to complete the road next year, and places himself on record by saying that his trains will be able to run to Tillamook Bay in less than two and a half hours from the heart of Portland.

A few days after this Sunday edition exclamation the paper added: "Mr. Potter has the reputation among his friends of never having touched an investment that has not forged to the front, an unqualified success . . . his observations are wonderfully discriminating. . . . He has enabled more people to make more money in suburban realty than any other individual in the United States."

In his Bay Ocean resort Potter recorded over three thousand 50-by-150-foot lots with several miles of streets and boulevards. Despite delays in building a rail line from Portland and an economic depression, lots sold easily at $500 to $1,800 each. Investors responded with capital for a grand hotel and a giant natatorium, or swimming palace, where visitors, after viewing the cold ocean, could swim in a grand pool of its heated waters. The grand hotel, three stories high, included bowling alleys, tennis courts, and a dance hall.

Meanwhile timber interests lobbied with Congress for a stable inlet through which their ships could pass between Tillamook Spit and the mainland's hills to the north. In 1917 the U. S. Army Corps of Engineers completed a mile-long jetty projecting seaward from the hills on the north side of the inlet. In 1932 the Corps repaired the jetty and extended it three hundred feet to catch sand that powerful ocean currents had been sweeping past the tip of the original jetty. The dunes on the spit, which had been retreating one foot a year, suddenly disappeared by six feet a year. That same year erosion undermined the sidewalk alongside the natatorium. Four years later the seaward side of the building collapsed and the hotel disappeared. Most of the residents of Bay Ocean had already abandoned the depressed resort. Those remaining pleaded with the Corps for a south jetty, hoping it would trap sand moving north and rebuild the spit. The Corps declined the invitation and blame, saying the problem was not the north jetty but a natural recession of the Oregon coast.

A January storm in 1939 sealed the fate of Bay Ocean. Huge waves battered the spit with boulders and tree trunks. After three days the spit was breached in two places, including the site of the natatorium. At Jackson Gap on the narrow south neck, the only road to Bay Ocean stopped at a fifty-foot-wide chasm. The town's waterpipe poked out into midair, pouring fresh water into the new inlet below. Mainland residents around Tillamook Bay began to fear the ocean would wipe out the spit altogether. The U. S. Army Corps of Engineers said all solutions were too expensive. Nevertheless, after the end of World War II speculators began buying the lots on which the county tax collector had foreclosed. The local *Headlight Herald* declared:

> The old belief that Bay Ocean has been carried out to sea is being exploded, and the more hardy souls who take time to look around and see what an enormous pile of sand is still to be washed away to eradicate Bay Ocean have come to the conclusion that it will be a long time if there is any further erosion at the rate that has been eroded in the past.

A year later, storms opened two new gaps in the spit. The largest, Rock Crusher Gap, was 250 feet across. Water service could only be restored by fire hoses hung in midair across the gap. The Pacific Ocean administered the *coup de grâce* in November 1952. This time storm waves destroyed four thousand feet of the beach ridge. At high tide an inlet three quarters of a mile wide connected the bay and ocean and left the resort a barrier island without road, electricity, or water.

By 1958 the quiet mainland coast was beginning to feel like an open-ocean beach. Log-booming operations were broken up by waves roaring through the gap. Levees along the tributary rivers were collapsing. A boat channel along the mainland shore was filling with sand. The same ocean sand covered three thousand acres of oyster beds. Tides leaping the spit no longer flushed the channel protected by the north jetty, and it began to silt in following the rules we explained for inlet areas along barrier island coasts.

In 1956, at a cost of almost $2 million, the U. S. Army Corps of Engineers built a wall of boulders across the back side of the gap, forcing the tides to use the inlet by the north jetty. Then in 1969 shoals began to block this inlet. Much of the sand in these shoals came from the erosion of the spit. The Corps decided to meet the problem with a south jetty extending seaward from the point of the spit. Here, thirty-five years later, was the jetty residents had felt would solve their original problems back in 1932. After discovering the new jetty would not reach beyond the shoals, the Corps built it out another 2,800 feet. The inlet is now protected by a rock corridor one mile long.

Behind the south jetty the beach is widening rapidly. In fact, beaches behind both the north and south jetties are building seaward very fast. Does this mean sand travels in both directions on this coast? It does, moving south in spring and summer and north in winter as wave direction changes. Geologist Paul Komar and Thomas Terich, of Oregon State University, have measured the shoals and the beach growth behind both jetties and conclude that the annual net sand transport is near zero. Why, then, did the north jetty cause the spit to erode?

Dr. Komar believes that the north jetty and sand behind it changed the depth of the inner continental shelf. He points to data from the Corps showing that south of the jetty the underwater slope has been getting steeper, bringing deepwater closer to the beach. Waves approach with less shoaling and energy loss and therefore grab more sand. The U. S. Army Corps of Engineers believes the sand accumulating behind the south jetty will rebuild the spit's beaches. Dr. Komar points out that both jetties weaken wave energy and reduce the currents necessary to carry the sand further down the beach.

Whoever is right, Dr. Komar's essential point is that even where there is no dominant direction to sand drift, jetties still cause a disturbance in the dynamic equilibrium that can wipe out enormous areas of naturally stable beach.

Nothing here claims that beaches which build up behind protective jetties have no value. On the Texas coast a portion of Galveston Island has grown almost a mile seaward since 1894, when a jetty was built 3.5 miles into the Gulf of Mexico to protect the entrance to Galveston Bay. On the new land the city has allowed developers to build new homes. These homes sit between the Gulf and the 21-foot-high seawall built to protect the rest of the city from storms like the 1900 hurricane that killed six thousand people. Ocean City, Maryland, has also added to its territory since the Corps of Engineers built jetties on both sides of the Ocean City Inlet—one at Ocean City, the other across the inlet on the northern end of Assateague Island. The Ocean City jetty trapped sand in the predominantly south-flowing currents, and Ocean City's island grew so much that by 1955 the beach had expanded almost to the end of the city's famous pier. But the value of this new land has to be weighed against certain costs. South of the Ocean City Inlet and downdrift from the jetties, the beautiful and undeveloped public seashore of Assateague Island has been slowly eroding. By 1962 the jetty on its side of the inlet was no longer connected to the shrinking northern end of the island. To keep the inlet from shoaling or finding a shortcut to the ocean through the island, the Assateague jetty had to be reconnected by dredged-up fill. To date, the island's shoreline has retreated some 1,750 feet.

Jetties may also serve development only to destroy it. The U. S. Army Corps of Engineers, for example, recently planned new jetties for Oregon Inlet on North Carolina's Outer Banks. The inlet's constant shifting has caused the partial collapse of the two-mile-long Bonner Bridge, which is the only road link between Hatteras Island and the mainland. Dr. Stan Riggs, a geologist from East Carolina University who predicted the bridge collapse, calculated the volume of water that can pass through the jetty corridor. During a hurricane, Riggs says, Oregon Inlet becomes a broad, deep channel, much wider than the space between the proposed jetties. This storm channel drains the floods from Pamlico Sound. Dr. Riggs predicts that in a major storm Pamlico Sound will break through a new inlet and that the most likely place is the heavily developed town of Nag's Head.

Breakwaters

The inlets protected by the numerous jetties of the East Coast service the harbors created by rising sea level flooding the broad, flat coastal plain behind barrier islands. On the West Coast, where both the edge of the continent and the continental shelf are very steep and natural harbors are scarce, people have had to confront the waves directly, building massive rock and concrete walls in the sea itself in order to protect ships and small boats from the ocean. In their constant confrontation with the nation's greatest waves, these breakwaters must be built taller and stronger than groins and jetties. We know of no wood or steel breakwaters in the open ocean. They are usually piles of boulders capped with concrete.

A successful breakwater creates behind its wall a calm energy shadow in which ships can moor in safety. This simple and bold confrontation of the ocean is often so successful that at times it silts in the very harbor it was built to protect.

The reason, of course, is that behind the breakwater waves do not contribute their drive to the longshore current. The current falters and sand drops out, creating a bulge in the beach. When the bulge begins to fill a boat harbor we have one of the rare instances of a growing beach becoming a liability. Santa Monica Bay, on the coast of the Los Angeles megalopolis, offers a classic example of breakwaters bestowing on their creators the desired harbors, and also a host of problems. Santa Monica Bay is rimmed by a long curving stretch of sandy beaches. At the north end of the Bay is Malibu. The southern stretch ends in the shale cliffs of Palos Verdes Hills. Glamorous and wealthy landowners hold title to most of the northern end of the bay. The southern portion is public, but crowding this narrow public strand is an almost unbroken string of luxury houses set on lots that often sell at over $100,000 for half an acre. (The public strand in front of private homes is convenient for those landowners, since the public pays to maintain a beach in front of their vulnerable oceanfront position.)

In 1933 Santa Monica answered its yachtsmen's call for a harbor by building a breakwater parallel to shore. Behind the breakwater was an elaborate recreational pier with many berths. Almost immediately the interrupted longshore currents began to drop their sand behind the breakwater in the "wave shadow." Within a year much of the new harbor was filled in. Downdrift the beach starved until it almost disappeared. Will Rogers commented that harbors like Santa Monica's would not be of much value unless we could learn to irrigate them.

The same kind of harbor and breakwater exist at Redondo Beach, California. When first built, the breakwater created a nice yacht basin, its elbow protecting boats from the big northwest waves. Hermosa Beach, north of the jetty, began to benefit by a broad beach building up behind the shore leg of the breakwater. To the south, Redondo

Beach began to slip away. The big waves first undermined the boardwalk. The U. S. Army Corps of Engineers quickly tried to shore it up with boulders, to no avail. The boardwalk fell in, and so did an entire waterfront block of buildings. In desperation the Corps extended the breakwater to the southeast and built a short jetty out from shore toward the tip of the new breakwater. South of the harbor the shore continues to slip away.

Many scientists and engineers have come forth with plans for ingenious variations on breakwaters, plans that would make them either less monolithic to allow sand to pass through, or more natural and therefore presumably more durable and esthetically more acceptable.

Most of the less monolithic breakwaters are walls with holes in them. Stacks of concrete pipes or blocks soak up wave energy as the water sloshes around in the holes. So, too, breakwaters armored with tetrapods or tripods—big concrete equivalents of a child's jacks—act like an energy sponge. These breakwaters (and groins and jetties) cut down wave reflection and swash that overtops the wall, but their energy shadows still waylay sand on its longshore journey. A number of permeable walls allow sand to pass through but only by sacrificing some of their protective function. Professor John Isaacs, of Scripps Institute of Oceanography, invented a very flexible breakwater after he watched a Coast Guard cutter trying to approach a buoy and noticed that the bobbing of the boat on the waves was slower than the buoy's elusive lurching. The buoy, he reasoned, must be absorbing wave energy. From this he invented a breakwater made from an anchored grid of five-foot-diameter steel spheres. The grid allows fish to pass freely, and tests show it should cut wave energy up to 75 per cent. The breakwater can be as large or as small or as dense as desired for fine-tuning wave control. It will be useful in many more locations than the standard rock wall, but that could be dangerous. Like any other breakwater, the field of steel balls interrupts the natural equilibrium. Nor has anyone yet seen hundreds of barnacle-coated five-foot-steel balls tempting swimmers and boaters. Broken loose in a major storm, they would be like cannonballs fed to the surf. Even if they stay anchored, a distressed boat might better run aground on a rock wall than drift into a field of thousands of these floating maces.

Across the continent in Florida, Mr. Edmund Boots, an 88-year-old retiree who is said to have invented the traffic light, claims to have the "ultimate solution" to stopping Florida's erosion. He noticed that Riomar Reef off Vero Beach was made not by slowly growing corals, but by the rapid sand-tube house-building of the little seballariid worm. Riomar Reef acts like a breakwater, cutting erosion behind it. In order to give the worm the rock base it needs to start a colony, Boots has invented a submerged concrete archway. The eight-foot-wide underwater worm boardwalk could be built for $1 million a mile, he says, and he has formed the Sabecon Reef Association with a retired rear admiral as administrator. While the reef would seem to be cheaper than a rock breakwater and capable of self-repair, it is hardly the "ultimate solution." Where natural worm reefs appear, the beach downdrift has had decades or even hundreds of years to adjust. Any new reef on a developed shoreline would inevitably cause downdrift erosion and property loss like any other breakwater. Mr. Boots's association must either be ready to introduce this can of worms along many miles of the Florida coast or not at all. And once established in quantity anywhere the worm and its natural rock habitat do not occur, someone is going to have to be responsible for introducing a whole new ecosystem with all its unknowns.

The list of exotic breakwaters, groins, and jetties is as long as the list of people who study the problem. Since nature is so adaptable and varied, so are her imitators. What her imitators cannot duplicate economically are the long-term readjustments that nature makes to all major alterations in the face of the beach. Every new beach defense would require suspending property lines and sacrificing development at least for several years while each shape and dominant wave patterns changed in response to the new structures.

The history of offshore structures certainly suggest strongly that the best defense in this instance is not a good offense, at least not one which takes us into the surf zone. It is both cheaper and safer to defend on land itself. So let us retreat from the surf and see what can be done on land or at least with its solid backing.

Seawalls, Bulkheads, and Revetments

A favorite protection for waterfront property is the bulkhead, a low wall of wood, steel, concrete, plastic, or fiberglass behind which a private lawn or motel patio remains flat and grassy—at least for a while. Bulkheads function as a replacement for dunes on the open beach for natural vegetation on quieter shores. A rigid wall, in other words, takes the place of the flexible natural system. In calm weather, from a lawn or patio, mankind and nature seem at peace, each holding proper territory. Yet any structure on the beach which retains sand in place, reduces the beach's ability to respond to the peaks of tide and storm energy. The normal response of a beach to a big storm is to yield sand from the dunes and allow the waves to draw it across the lower beach. As this happens the beach slope becomes flatter, the waves encounter more friction in their headlong rush, and the beach survives. Sand from offshore bars washes back onto the beach and blows into the rebuilding dunes.

Where the beach cannot flatten itself, waves carry off more sand and drop it in deeper water or in locations down the beach from which it will not return. Because property owners feel that a bulkhead invites the public to see it as a boundary line, they usually place the wall at the high-tide line or even below, claiming some of the public domain. The closer a bulkhead is to the water, the quicker the beach will disappear. As storms are encouraged to steal the beaches from in front of bulkheads, groins will appear to pull back sand. When this fails, sand pumping or beach nourishment is tried. This continuing expense usually becomes too burdensome for a community, or else occasional storm overwash makes the new sand seem purely cosmetic. Sooner or later the bulkhead is replaced by a more massive seawall that might survive without a beach.

Seawalls are the ultimate fate of most developed beaches. A major problem with seawalls was summed up by the bemused observation made to us by the building inspector at Monmouth Beach, New Jersey. He said it appeared to him that every year the waves were getting larger and more violent. Storms seemed to cause more damage. No one has measured enough New Jersey waves to verify his observation, but he should be right. In his day he has seen not only houses but farms disappear where the sea lies in front of the seawall. Seawalls, like bulkheads, lock up sand needed to maintain the beach, and they reflect wave energy. Being below the high-tide line, their effect is more constant than a bulkhead's. The beach profile becomes steeper, waves encounter less friction and pound the sand with more force, both direct and reflected. This effect surmised for Monmouth Beach, has been documented by the U. S. Army Corps of Engineers in front of Miami Beach.

Long Island and New England beaches and some West Coast beaches are special cases. Here property owners try to stop cliff and bluff erosion with seawalls. In the process they cut off supplies of the very materials of which the beach is made. Few of these cliff-front seawalls last more than a decade before the disappearance of the beach undermines them and forces their repair or reconstruction. In some cases seawalls cost more than the lots, or even the houses, they protect.

When water does not undermine a seawall from the front, it is likely to attack from behind. Waves striking a flat seawall slop over the top and return to the sea through the foundation. The collapse of many walls is foretold by gaping holes opening on the landward side. This slopover can be controlled by a wall that curves seaward at the top and throws the waves back on themselves. The cost of such a wall is often more than most communities can bear.

All seawalls are expensive, but sometimes the comfort they seem to offer is irresistible. Many American communities, through boldness, extravagance, or necessity, have bypassed bulkheads and groins altogether and erected massive seawalls. Galveston did this after it became, in 1900, the first American city to be destroyed by a hurricane. Its 21-foot-high seawall has successfully turned back several hurricanes, but the beach has been sacrificed. Only at the north end of the island where sand has piled up against the jetty has the beach held its own or grown. Where three hundred feet of beach once lay in front of the wall, today before most of the wall only a narrow skirt of rubble protects the wall's foundations from scouring. The west end of the wall has had to be repeatedly reinforced with expensive concrete aprons. Almost every seawall in the nation has created the same problem. The advent of the wall transforms the active recreation of swimming and beachcombing into the passive recreation of looking over a wall—if it's not too high.

The state of Louisiana has a unique reason for proposing the nation's largest ever shoreline armoring project. The state has already spent several million dollars trying to stabilize the shoreline of the Mississippi Delta and its rapidly retreating islands. At stake are not towns or resorts but millions of dollars in revenues obtained from oil wells within the state's three-mile marine jurisdiction. In some cases the three-mile limit is measured from narrow strips of marsh along the edges of huge ponds or bogs at the margins of the delta. If these marsh strips erode, the ponds become part of the Gulf of Mexico and the legal shoreline may leap backward as much as thirty to forty miles. This would leave many oil wells out in federal territory.

So far the revenue rescue operation has only produced revetments on Isle Derniere and West Timbalier Island. The state, however, is thinking about using federal Coastal Energy Impact Funds to begin more work on its massive plan. These funds, originally intended in part to help alleviate environmental damage, could in fact multiply it. A ray of hope shines in the words of a Corps of Engineers official who told us, "This is an impossible project."

Lest we appear to have a limited perspective on the cost of seawalls, we should mention the cheapest, quickest one we've heard of. In La Jolla, California, at the height of the 1978 winter storm, a homeowner called on an engineering firm for immediate help. Representatives of the company soon appeared on the beach driving the cheapest vehicles they could buy on the local used-car lots. They parked their cars in front of the threatened home. The instant seawall worked and its pieces were later towed away to the crusher, leaving behind a satisfied homeowner. But he still needed a beach.

Leaving out such temporary solutions, perhaps the cheapest alternative to the seawall and bulkhead is the revetment. This is merely a wall tilted back away from the waves and so imitating, as well as stone and cement can, the way the natural beach flattens out under wave attack. The revetment usually lies upon the foot of the first dune. Requiring no poured foundation or form work, it costs half as much as a seawall and is easier to build than a bulkhead. If built of boulders, rubble, or concrete blocks, it has the added advantage of soaking up wave energy. Revetments, of course, are not as convenient as seawalls and bulkheads —not as easy to fish from, tie a boat to, or even to sit on. Those that imitate nature best, rock and rubble, are least convenient of all.

To many communities a revetment more than pays its way. Only Ediz Hook, a slender three-mile-long spit, protects Port Angeles, Washington, and its busy harbor from the rough water of the strait of San Juan de Fuca. A revetment saved the spit when it began to erode badly in 1930. Since then the revetment has grown to eighteen feet high and over two miles long. The U. S. Army Corps of Engineers considers it a model project. The real story, however, lies in the cause of the erosions—a 72-inch waterline built along the base of glacial bluffs in 1930 and protected by a bulkhead. The bluffs could no longer yield the material to maintain Ediz Hook and protect the harbor. The Corps told us that the fresh-water pipes could be rerouted along the top of the bluffs more cheaply than maintaining the bulkheads, armoring Ediz Hook and pumping new sand onto it every year. Much of the property on top of the bluff, however, is owned by local government officials and administrators of the Crown Zellerbach paper mill, a major water user. Local sources say politics have caused Ediz Hook to be armored at a cost of over $6 million. Every year the spit needs new sand to replace the beach lost in front of the revetment.

The connection between revetments, seawalls, bulkheads, and the eventual need for beach nourishment is the same at Ediz Hook as at Cape May, Atlantic City, and Miami Beach. So the chain of events comes full circle. First a natural beach protects itself. Then, as property lines are staked out, groins and jetties appear. Then bulkheads and next seawalls. Finally, to protect the seawall and the local economy, the beach is recreated. The difference is that in the beginning no one bore any cost. In the end a highly complex technology becomes an unending expense.

Replenishment and Dune Building

If nature can build a beach out of sand, why can't people? Why not simply shape sand into beaches and dunes the way nature shapes a beach? We can indeed build a beach by simply dumping sand at the water's edge. Likewise, we can build dunes. All over the Atlantic seaboard, from Green Pond on Rhode Island to Key Biscayne, homeowners and local governments are putting up sand fences and planting beach grasses and laying down barriers of scrap lumber and old Christmas trees to trap sand and rebuild the dunes bulldozed or eroded as a result of development. But our beaches and dunes, like nature's, will move to preserve themselves, and that is the very problem engineers are trying to solve. When we build a beach we want it to stay in the shape we built it. Although a fixed shape and a healthy beach are contradictions, not long ago everyone agreed that beaches could be "naturally" fixed in place. In the 1930s the federal government launched the most colossal dune-building project in history. Geologists, engineers, and politicians agreed that by using the Civilian Conservation Corps they could fight both the Depression and the sea. No one doubted, at that time, that shoreline erosion would soon destroy the Outer Banks. Most scientists and historians believed that the restlessness of the Outer Banks had only begun when colonists stripped the islands of timber and turned their goats and cattle loose on the stabilizing dune grasses. The idea of righting the wrongs of history and of hundreds of workers saving a spectacular landscape had great psychological and economic appeal in the Depression. Along over thirty miles of beach the CCC planted millions of trees and shrubs and plugs of beach grass. They staked out hundreds of miles of slatted sand fence. The aim was to trap sand and build a high wall of foredunes. Before long, scientists and government officials looked at the fifteen- to twenty-foot-high dunes and felt they had saved the threatened islands. Now, some forty years later, we have discovered that the beautiful dunes are not a replica of the pre-colonial natural system and are actually destroying the very islands they are supposed to protect.

Dr. Robert Dolan, of the University of Virginia, who has studied cross sections of the islands and innumerable maps and charts, finds that the dunes have acted much like a seawall. Because they are too high to permit overtopping and too continuous to allow inlets and breakthroughs, except under extreme conditions, the ocean's energy has been concentrated on the beaches, most of which form Cape Hatteras National Seashore. The beaches have narrowed and the offshore profile is growing steeper, creating stronger waves. The dunes themselves are so successfully planted that both their back and front sides are very steep. Waves strike this steep face with greater impact than a gentle slope, and storm erosion is fast and spectacular. The protection the dunes first offered seems to have lasted just long enough to attract enough development behind them for a major disaster.

Homes, motels, and shopping areas have moved up behind the dunes as confidently as behind the Galveston seawall. Dr. Jack Pierce, a geologist with the Smithsonian Institution, says the dune wall, while intact, acts as a dike damming up flood waters rising in Pamlico Sound behind Hatteras Island during a major storm. He suggests that a strong dune line may leave the towns behind it as the only weak points. This may be especially true since the high dune wall catches sand which would normally blow inland over lower, less completely vegetated dunes to form secondary dunes and a higher midsection for the island. Paul Godfrey, of the National Park Service, has discovered another threat posed by the dune wall. Because the wall stops all overwash, the marshes on the back side of the islands, starved for new sand, gradually erode, making the islands narrower and more likely to be breached in a large storm. Winter storms of 1973 suggest how close we might be to witnessing the truth of Godfrey and Pierce's analyses. High waves topped some dunes and carved through the steep narrow walls of others. Roads behind the dunes were flooded (this time from the ocean, not the sound), and in several places inlets threatened to slice through the island, especially the resort town of Buxton, North Carolina where several motels and houses collapsed. Only highway department bulldozers pushing sand back into the dunes around the clock saved Buxton from being the first town destroyed by a protective dune wall.

When the man-made dunes began to yield, the Corps' response was to pump new sand onto the beaches. In order to protect the dune, an expensive artificial beach was necessary. Where erosion threatened Cape Hatteras Lighthouse, the artificial beach was held in place with groins. Of course, just south of the lighthouse the beach and the dunes began to erode as the groins cut off the longshore current.

Simple sand replenishment would seem to be the most natural of all engineering solutions. We can find little environmental fault in the idea of putting sand on top of sand, but the idea is much simpler than the fact. The biggest problem is often where to find the sand.

The traditional solution is to pump sand out of the bay across the island onto the open beaches. With our new admiration for the productivity of marshes and estuaries, we have become increasingly reluctant to diminish their value by stirring up massive amounts of mud. Also, a hole in a lagoon will change wave patterns and energy, causing erosion on the back side of the island where none previously existed. Finally, the lagoon gathers its sand under conditions much calmer than the ocean. The sand is finer than beach sand and often disappears rapidly when exposed to ocean surf.

The dredgers have been forced out of the bays and lagoons onto the continental shelf, where dredging is much more expensive. And we are still stuck with a hole that may either increase wave energy or change refraction patterns, or both. The U. S. Army Corps of Engineers is trying a compromise between lagoons and shelf by using the sand dredged during routine maintenance of inlets and channels. This is certainly less wasteful than traditional hopper dredging, which simply barges the sand out to deepwater, where it is lost to the nearshore system, at least until sea level falls many thousands of years hence. The fly in the environmental ointment of using inlet and channel sand is that near major harbors industrial wastes of all kinds are bound up in these sands, and the beaches are hardly a good disposal site.

In the end, replenishment has a much more fatal flaw than sand size or pollution. For practical purposes it is nothing more than cosmetics. The true beach, as we have said earlier, is more than the bathing strand. It is a wedge of sediment three or four miles wide stretching underwater to depths of thirty or forty feet. Replenishment drops sand only on the thin visible strip of upper beach. For obvious reasons, no one has yet suggested building up the entire shoreface down to thirty feet below the surface of the sea.

The net effect of replenishing only the upper beach is to steepen the beach profile. The beach wants to return to its natural, more uniform shape. The steeper profile of replenished beaches is the reason they erode more rapidly relative to the natural beach. To maintain its shape a replenished beach has to be re-replenished and re-re-replenished. As nearby sand supplies disappear, more and more expensive pipe and boosters separate the dredge from the beach.

The public is so relieved and awed to see a broad new replenished beach that the temporary nature of the solution is never recognized. A re-replenished beach disappeared at North Carolina's Wrightsville Beach in 1978, leaving only a line of steep man-made dunes facing the sea. To make a more useful beach city fathers bulldozed the front part of the dunes and spread the sand out as a new beach. The offshore slope had already become unnaturally steep, and 1978 storms wiped out the new beach, leaving a five- to six-foot scarp between the beach and the first dunes. Small children now have to be lowered to the beach by parents.

The great dune wall of the Outer Banks shares much in

common with Richard Sylvester and Siew-Koon Ho's proposal to save the National Seashore with a façade of artificial headlands. Since the unanimous endorsement in the 1930s of the dune wall, coastal engineers and shoreline geologists have gone their separate ways.

Engineers have increasingly refined their calculations of stress and strain both in the natural system and in the structures they build to defend the beach. They can reduce wave and wind forces, weather probabilities, and the strength of their own materials to formulas and precise numbers. Computers can digest the numbers and design new defenses. This new sophistication has yielded a supermarket of new designs, many of which imitate nature in some way. Experts say at least two hundred designs for "low cost" protection were published between 1974 and 1978. Professor Billy Edge, of Clemson University, a member of the National Shoreline Erosion Demonstration Program, says only 10 per cent of the low-cost solutions show any promise. Among these he lists artificial seaweed developed in Germany, concrete blocks preglued to a filter-cloth mat, cement bags, a reef of buoyant frisbees, and several kinds of sandbags. He still reserves judgment on all of these devices, now being monitored by the Demonstration Program. One judgment we can make is that however novel, all have the same intent—substituting a status quo amidst natural changes. Edge holds up the simplest solution with least reserve. He points out that a very effective dune-preservation measure was proposed by the Nassau–Suffolk Regional Planning Commission: plant the dunes with poison ivy.

Perhaps because engineers have had so little success in taming the sea, they are quick to repeat whatever seems to have worked. Thus, Sylvester and Ho leap from Singapore to North Carolina with their artificial headlands. Geologists, whose job more often is to describe than to build walls or solve problems, have become increasingly aware of beaches as whole systems and of the differences from one beach to the next. Wave conditions, abundance and quality of materials, rate of sea-level rise, and shoreline shape are quite different in Singapore and North Carolina. Even beaches on adjacent islands of the Outer banks differ more than most sets of fraternal twins. Take the case of two islands we described in the Cape Lookout National seashore —Core Banks and Shackleford Island. Both islands share the same plentiful sand supply, but dominant wave direction and longshore currents are very different. The wind blows up or down the length of Core Banks, and sand deposited on the beach by the waves is not blown inland to form dunes. In a storm the island's simple response is to flatten out and endure complete flooding. The wind blowing onto Shackleford from the south pushes sand into extensive dunes which prevent most flooding. The dunes are a reservoir of sand to flatten the beach profile during storms. A seawall here, locking up this sand, would do more damage than on Core Banks.

Because engineers are called upon to treat specific and local symptoms, their field of vision seldom includes the full scope of beach movement. Most beaches are larger than the problems treated by engineers. The shoreface extends up to four miles offshore into water thirty feet deep. On barrier islands all visible land and all development are part of the dynamic equilibrium. While whole islands do not experience the rapid changes seen on their beaches, their evolution is not nearly as slow as the evolution of an animal species. Within a human lifetime we can see the hands of the clock move. What makes island evolution predictable and gradual is the amazing flexibility of the beach itself. When engineering changes the character of the beach and tries to stabilize it, unpredictable change accelerates on the whole island. Wave attack may become stronger and storms more destructive. Mid-island dunefields may begin to disappear. The marshes on the lagoon side shrink. Fish and shellfish grounds disappear, wildlife populations drop. Polluted water running off the mainland or seeping out of septic fields is no longer purified by the natural filter system. The island becomes a city at sea with the water rising slowly around it.

The difference between the holistic perspective of the geologist and the problem-focused vision of the engineer, like all conceptual differences, affects their language. Between the words *erosion* and *migration* lies the gulf that now separates the two disciplines. To property owners and to engineers employed by them, when a shoreline moves into the boundaries of any piece of property, land is eroding. The word carries with it the sense of irreparable loss of vital land, images of the Dust Bowl and gullies on worn-out southern farms. Within the world of private property, moving shorelines are trespassers to be subdued and evicted. Geologists, looking at the larger picture in time and space, see no permanent loss but a migration, islands rolling over themselves in tank-tread fashion. Migration, whether for birds or beaches, is the process of travel for the sake of survival.

Since both words denote concrete and measurable changes, which best describes the beaches? Property is indeed lost as beaches move, and sand is taken away here and added there just as inland erosion moves soil from one ecosystem to another, from field to river or swamp, or from swamp to ocean. The moving sand of the beaches stays in the beach system or is replaced by sand arriving from fresh sources. The beach ecosystem, under natural conditions, does not lose any vital material; it merely moves and survives. Migration best describes the natural beach.

Actual loss of sand and whole beaches, and even whole

islands, does occur, mainly as a result of human interference. Groins, jetties, breakwaters, dams, seawalls, bulkheads, and even artificial dunes, prohibit the beach from using its natural defenses. As with the Dust Bowl and gullied farmlands, serious and rapid erosion is most often caused by man, not defeated by him.

This is not to say that island migration poses no problem for mankind. It burdens our coast with problems whose weight is in proportion to coastal development. So why quibble over words? Why not defend valuable development against migration? Because migration is itself a survival mechanism, a response to a great force. To replace migration with engineering we must ask if our work meets the natural adversary as economically and durably as migration. Geologists say the adversary is nothing less than melting polar icecaps and rising sea level. When Richard Silvester's critics said his approach did not figure on the consequences of rising sea level, he replied: "The subject of . . . sea level hardly warrants a comment." Any problems caused by rising sea level, he asserts, "will be solved by future generations." This response from a highly respected engineer is no more sophisticated than the comments we mentioned earlier by a Virginia Beach reporter, who calculated that folks had better hurry up and be prepared to evacuate by the year A.D. 4500.

As sea level rises, islands and beaches do not stand still and allow water to pass over them. Like a well-disciplined retreating army, they move back through a series of complex maneuvers. These include inlet formation during storms, inlet tidal deltas, inlet migration, overwash, and dune formation. Engineers who also know geology estimate that for every one foot rise in sea level the beach retreats anywhere from a hundred to a thousand feet. Some islands in South Carolina are rolling over themselves in retreat so fast that salt-marsh grasses are still anchored in the mud that reappears on the ocean side. In New England sea level now seems to be rising at the rapid rate of three feet a century. Where the land is sinking, sea level rises even faster.

We are not entirely happy as we write what amounts to an accusation of engineering ignorance, although we think the evidence is ample. Nevertheless we are happy to have some admissions of ignorance from the engineering community. David Bella, professor of civil engineering at Oregon State University, says coastal engineers are "accustomed to working in ignorance and have developed a formal factor to weave it into their equations." This is none other than the "factor of safety," extra strength or size built into a project which should absorb any forces unknown to the engineer or designer. Whatever the built-in safety or ignorance factor might be, it exists mainly for the safety of development, not for the protection of the beach. And whether stated or not, so long as the beach is destroyed, all safety factors have a time limit.

Coastal engineering will grow more sophisticated and more inventive. Neither will solve our problems so long as shoreline engineering remains the extension of a terrestrial mind into a world that is undeniably ruled by the oceans and rising sea level. Technology is a set of tools to increase mankind's durability, stability, and strength. This is true of plows, cars, fiberglass insulation, the Concorde SST and seawalls. On the beach technology has served to move us closer and closer to the waterline and to the still mysterious power of the oceans. As on a battlefield, the closer we move toward that power, the greater the power we need for defense and the greater the potential for catastrophe. In a similar situation on the moon, we have admitted the ephemeral nature of our support and defense systems. But then, masses of voters have not yet frolicked in the moon's craters or purchased land on its mountains or dry seas. Most of the beaches are staked out, and we insist that if our present defenses fall, we can and must build more lasting ones.

King Canute, being a monarch, could knock his subjects' faith in him by commanding the waves to retreat. The public still wants the sea and not the beaches to retreat. Local, state, and federal governments, obedient public servants, still do the master's bidding. The masters, however, are beginning to fight among themselves, a complex legal struggle in which individuals fight not only each other, but fight themselves in their single role as the American public. We will only know the significance of this internal fighting for the future of the beaches if we study the complex rule of coastal law that is emerging to keep the combatants from drowning each other. It is still an open question whether this law will or will not recognize the natural system being fought over.

11

Who Owns the Beaches?

THE LAW OF private property is usually clear: you stake out and own a piece of land. Once upon a time rocks, trees, rivers, ridges, and stone walls were called upon to witness where one person's territory stopped and another's began. Today iron stakes and concrete monuments take their place, all tied into imaginary but unmoving lines of latitude and longitude. Everything within the boundaries of your property is yours unless you sign it away. Similarly an automobile is private property and everything from the wax finish inward is yours in all fifty states, no matter whether the car is in your driveway, on the Interstate, or, by some drunken misfortune, at the bottom of Niagara Falls. But is the law so clear about a piece of land that travels?

Here law becomes confusing. When the United States gained independence the land below the high-tide line belonged to the king, who held it as trustee for the people. The king was replaced by the states, some of whom try to sell or give away the land. Do they have the right to convey it? If this is public trust land, does the public have the right to get to it? If tidelands belong to the public, why do so many towns reserve beaches for its residents only? Since the tide's reach differs every day and differs drastically by the season, where do the tidelands begin?

Coastal law is the chaotic battlefield on which our firm and orderly notion of private property and real estate battles with the huge dynamic forces of nature, which recognize legal systems even less than iron or concrete markers set in shifting sands. Because the legal system has never adapted to the surprises of nature, people who rely on the law for peace among people have also met disappointment. The public's interest in and love for the tidelands gives all of us a stake in what happens anywhere on the beach. Like neighbors who individually own no significant lot, we have both an emotional and legal interest in the entire neighborhood.

Unfortunately, as the beaches were transformed into real estate bonanzas, government stood back. As the nation developed, the public interest in the beaches suffered irreversibly. Only 7 per cent of our beaches in the lower forty-eight states are available for public recreation, and almost all of these are overcrowded. Half of all Americans live within fifty miles of the beach. Over 75 per cent of the nation's growth occurs in this crowded border. While demand for public space grows faster than the population, demand for private beaches is so great that Americans, wanting second-home lots and condominiums near the water, will pay from $1,000 to $30,000 for a foot of shoreline.

Given the great importance of property to our emotional and economic lives, the defense of it is stubborn, fierce, and extravagant. At the beach it is defended first of all against its original creator and final possessor, Nature—and the results we have seen. Second, property—at least private property—is defended against other people. As beaches physically disappear, what remains becomes increasingly fortified by walls, fences, gates, and NO TRESPASSING signs. Often it even becomes less visible. On Florida's Atlantic coast Broward County's Galt Ocean Mile, a line of hotels, motels, and apartments, is known to residents as "the Galt Wall." On the Gulf Coast islands, where thousands of retirees with modest means live in cottages and trailers, both their view of and access to the Gulf waters have often been cut off by a solid line of high-rise motels and condominiums. The beaches themselves, where the retirees once spent hours swimming, fishing, sunning, and gathering shells, disappeared in front of seawalls and bulkheads.

It is not simply the desire for private property that creates conflict between the general public and private landholders. The technology used to defend the private domain often succeeds by destroying the public domain. No beach problem exists until lot lines are set out and buildings constructed. In order to protect this unusually vulnerable private property, often owned by unusually powerful citizens, Congress and the states have responded to beach problems with special laws to pay for expensive technology. These laws have encouraged development and massive engineering. Protecting beach property at first seemed to be a proper government function. As in other cases where a group of people cannot afford to pay for and protect their own welfare, the government has passed legislation that seeks to protect private property in the same way it protects the nutrition and shelter of poor families.

Property owners on the ocean and gulf beaches have benefited not so much from their own efforts as from the efforts of Americans in the Midwest. The appearance of shoreline legislation often waxes and wanes with the levels of the Great Lakes. Some of the largest shoreline cities and the greatest wealth and number of Congressmen are in the Great Lakes area—Milwaukee, Cleveland, Detroit, etc. And here many shoreline problems, in particular sea or lake level, are a more immediate threat to private property. The level of the Great Lakes since 1860 was downward. But from 1965 to the present the level has been rising. Since 1969, Congressmen from the Great Lakes states have been applying increasing pressure on Congress and the U. S. Army Corps of Engineers for help. Emergency legislation to aid some 214 miles of critically eroding lakefront in cities and towns around the Great Lakes sets the stage for an engineered seacoast where engineering is often unneeded.

In making a plea for Congress to protect private property along the shore and provide up to 90 per cent of the cost of erosion control (current top limit is 70 per cent), Representative Henry Shadeberg of Wisconsin in 1970 justified greater federal expenditures by arguing that the protection of private property would maintain the local tax base and keep real estate values from declining. Representative William Minshall of Ohio said: "Those of you familiar with our voting records know that Congressman Shadeberg and I are not known to be big spenders of taxpayers' money. We do not on principle encourage citizens to run to Washington for funds until every resource of private, municipal, and state endeavor has been exhausted. That is the case in this instance." Many lakefront residents soon had greater reason than ever to wish Congress had gone with Minshall and Shadeberg.

Lake levels in the spring of 1973 reached historic highs, pushing damages close to the $1 billion mark. Simultaneously legislation appeared that would have committed unprecedented amounts of public money to protect private beaches. Illinois' Senator Adlai Stevenson, III, rallied coastal state senators behind a package of bills which would have been an economic boon to shoreline property owners. One would have reimbursed residents for damaging effects of federal projects (e.g., erosion downdrift from a jetty). A much more far-reaching bill would have allowed the government and the Corps to provide direct aid to save private property threatened by shoreline changes.

Senator Stevenson had trouble lining up scientists to speak in favor of the bills. Most expert witnesses spoke with conviction only of a bill committing money for shoreline studies. The most crippling opposition to aiding private landowners came from the U. S. Army Corps of Engineers itself, which found none of the bills acceptable. This is a case where Corps' critics must admit that the Corps was throwing out some projects that might have added to its authority and scope. The bills never came to life after the hearings.

Although the rate of beach development has been awesome, especially in the past twenty years, public officials have had to be delicate in their efforts to gain public support to aid and maintain private development. The fate of Senator Stevenson's bills indicates that legislating support for private development will become even more delicate as time passes—for several reasons. First, the public, which has few beaches of its own, increasingly rebels at allowing future sites to be locked up by development, and at spending public money to protect a scarce resource for only a few people. Underlying this rebellion are even greater political, economic, and social questions.

Each year disaster aid becomes a larger item in the federal budget. It is not nearly as large as defense or welfare expenditures, but it is called for loudly and dramatically, and it has seldom been refused. The public is beginning to see that on the beaches the stage is clearly being set for major disasters in the near future. Engineer Lim Vallianos, of the Corps, says it is a tragedy that severe hurricanes have not struck the Atlantic seaboard full blast during the last decade. The result, he says, is that developers have taken full advantage of a strong sense of complacency. When a hurricane the size of Camille strikes Miami Beach or Long Island's south shore, the damages will be in the hundreds of millions, perhaps over a billion dollars.

Some people have justified shoreline protection expenditures by comparing the situation to that in the flood plains of rivers. But in protecting the flood plains at least some argument can be made for saving farms. No high national priority is at stake in the protection of hotels, motels, and beach cottages. Even if it were, the beaches differ from flood plains in that rising sea level assures continued flooding and erosion even if no storm is ever born again in the open sea.

Beach property owners are not only faced with local and federal government growing more and more reluctant to protect them against nature, but the wave of populism that marked the 1960s has created an overwhelming political pressure for public access to every shoreline.

In California, where new developments must include beach-access corridors, older residents have raised a great cry against letting the public on the beach. A private lot is indeed devalued when the public has easy access to the beach in front of it. The more exclusive the waterfront neighborhood, the farther its values fall when the public begins to walk, surf, swim, and fish in front of the picture windows. Passing fishing boats, freighters, and tugs are part of the scene, but working people simply enjoying themselves is not only a social equalizer but a value equalizer. However practical and businesslike the shore dwellers were when they accumulated their wealth, and no matter how enmeshed they are with trade and commerce, their reason for being on the beach is to enjoy the unspoiled vista, the fresh breezes, the crash of the waves, a solitary walk along the morning or evening beach.

Even those of us who cannot afford such unspoiled closeness to nature, hang on our walls scenes of wild dunes, green waves, and moonlit oceans. Few beach pictures are full of tourists, shell collectors, tanned surfers, or families sprawled on their blankets. No doubt about it, anyone will pay more for exclusive beach property than for a house fronting a public beach. As noble as the motives are, when a town, county, or state purchases a footpath or parking lot for the public to reach the water, nearby property values go down, or at least they no longer rise with such extraordinary speed. But no one should argue against public access on the basis of a diminished tax base.

On an inaccessible beach the value of the environment accrues solely to the adjacent private property. When the public gains access, the value lost by shorefront landowners reappears, many times multiplied, as a benefit attached to every home in the community. When everyone in town can get to the beach, the town becomes a more desirable place to live and property values rise on every block. We might say that providing public access is a way of redistributing wealth, but this assumes the wealth once belonged to the shorefront property. Given the ancient public claim to the tidelands, the rivers, bays, and oceans, it may be argued that the person who sells a shorefront home or motel pockets a profit which is largely created by the public domain. Certainly the dunes and the land immediately behind them were considered barren and useless until we learned to bulldoze house sites, build on stilts, deliver potable water, and control drifting sand. And still, it is not the land itself that is of so much value as its location next to the waterline.

The question of public access has taken on many features of a war between adjacent property owners. If we divide the shore into three bands, we can easily see the shape of the battle. The law recognizes three beaches—the wet sand or tidelands, the dry sand beach from mean high water to the vegetation line, and the uplands which lie landward of the dunes or between the dunes and the nearest road. The law also recognizes as a separate realm the sea and the seafloor beyond the low-tide line. The fact that these four areas of "beach" are part of one dynamic system and even part of a geologically unified wedge of sand moving up the coastal plain, has not yet been recognized by the courts or by lawmakers. Separate judicial precedents and statutes govern each band of beach as if they were distinct territories. Below the low-water mark the law seems most settled and least vulnerable to challenge. States generally own the seafloor three miles from the shoreline, while the federal government has broad constitutional rights to patrol the waters and regulate all traffic and commerce. No one has ever raised a serious question of private property in this realm, and even the giant oil companies line up obediently to bid for offshore drilling rights.

Although six states, by virtue of colonial ordinances, recognize private-property boundaries at the low-tide line, the public claim to tidelands is well-established on the greatest stretch of American coastline. The history of the public's claim is rooted in the same period of history that saw Western civilization begin to battle nature with breakwaters, jetties, and dredges. Despite Romans' willingness to reshape the shoreline, Roman law declared both the sea and its shores to be *res communes*, things of common use by all citizens, while the sea and its fish belonged to all nations. The maritime free-trade zone preserved the trade routes and fishing industry which were so important to the empire. The wisdom of this approach may be proven by the contrast of the early Middle Ages, when feudal lords and cities tore up the law and laid claim to large stretches of coast and to the shipping lanes nearby. Often local lords gained their title as a favor from the king. In those years of economic trouble and political despotism few were bold enough to challenge the king. Yet some men never forgot whose lands the king was giving away.

When King John signed the Magna Carta in 1215 he conceded any right he had to give away or sell the tidelands. As English common law developed, it became clear that while the king still held title to the tidelands and he could make grants of them to his favorites, his title was burdened with the public trust. Common law wrote on all the grants and deeds made by the king the overriding rights of the general public, rights which were to remain a kind of invisible ink on all conveyances of tidelands, even in the New World. This invisible trust was clearly read and declared by a wise judiciary over several centuries. The accumulated decisions of many courts gave birth to the *public trust doctrine.* Its existence was not in question when the American colonies sought their independence.

Perhaps the American public is fortunate that the colonists had no great interest in claiming the beaches and tidelands. If they had, the Founding Fathers in their respect for private property might well have cast aside the public trust doctrine of common law. Since they did not consider the question of tidelands or beaches, common law became, de facto, the law of the thirteen colonies and of the states which followed them into the union later.

The states in some cases have been able to shuck this political burden by special legislation or by virtue of colonial ordinances which modified common law. However, only six of twenty-three coastal states—Delaware, Massachusetts, Maine, New Hampshire, Pennsylvania, and Virginia—have extended private ownership to the low-tide line. Where the tidelands have been conveyed by legislation after statehood, private ownership may not be as firm as it seems. Until 1978, Georgia's waterfront owners thought their property extended to the low-tide line by virtue of a nineteenth-century act of the legislature. When the court examined the law in 1977 it found that private landowners had only been given the right to harvest oysters and clams.

In all states the federal government retains the right to regulate construction and dredging and filling to protect the public interest. This regulatory power and the public claim is overlaid with the constitutional mandate that the federal government shall have the right to protect and regulate commerce in navigable waters. Between the states and the federal government the public interest in the tidelands and beaches should have been well protected. In a new nation growing through a system of free enterprise and investment in private property for almost a hundred years, neither local, state, nor federal government made any attempt to protect what the law often called "wastelands." In fact, the first strong reaffirmation of the public trust doctrine by the U. S. Supreme Court was occasioned by a state legislature's attempt to give away submerged lands. The case which chiseled the public trust doctrine most clearly into American judicial history arose when in 1869 the Illinois legislature granted land under Chicago Harbor to the Illinois Central Railroad Company. The Supreme Court declared the legislature "incompetent" to convey the submerged lands. Justice Field summed up by quoting a New Jersey case: "The sovereign power, itself, therefore, cannot, consistent with the principles of the law of nature and the constitution of a well ordered society, make a direct and absolute grant of the waters of the state, divesting all citizens of their common right."

In 1894 the Court also declared that Congress was restrained by the same limits as state legislatures in matters of public trust lands. The case (Shively *v.* Bowlby) arose when a landowner claimed the right to mine sand near the mouth of the Columbia River, saying that his right to the submerged land had been established by a Congressional grant. In ruling against the landowner the Court relied heavily on famous treatises on the public trust doctrine written over a century ago by England's Lord Hale. While Hale's work described the king's role and powers, the Supreme Court held that the United States and the individual states were in essence filling the king's shoes. The Court affirmed that government cannot remove the blanket of public interest. The Court also reached back fifty years to one of its own decisions (Martin *v.* Waddell) and the declaration of Chief Justice Taney, who held that the public trust lands were necessary to the founding of the Republic: ". . . for the men who first formed English settlements could not have been expected to encounter the many hardships that unavoidably attended their emigration to the New World and to people the banks of its bays and rivers, if the land under the water at their very doors was liable to immediate appropriation by another as private property." Taney allowed that the king had granted whole colonies to individual proprietors, but he said that the river bottoms, tidelands, and seabeds were conveyed only as "a trust for the common use of the new community about to be established."

The United States began life with this legal heritage intact, and all Americans, no matter how poor or how far inland, possess an unshakable claim to the nation's most valuable and beautiful real estate. It is as if great-great-grandfather and -grandmother have left us a partial interest in a rapidly appreciating and highly desirable property. Our interest is not marketable unless a great majority of us inheritors agree to sell. The courts have also expanded the basis of the public claim as the public's interest has broadened.

Ironically, the strongest statement of the scope of the public interest came in the settlement of a dispute between owners of adjacent private beachfront properties. The dispute arose between Marks and Whitney, two neighbors on California's Tomales Bay. Marks, a developer, claimed he not only owned a big parcel adjacent to Whitney, but that his deed also gave him title to the tidelands in front of Whitney's lot by virtue of an 1874 patent issued by the governor. To make matters worse, Marks planned to fill and develop the land in front of Whitney's house and underneath Whitney's small pier. Whitney sued, saying his rights as both property owner and member of the public were about to be violated. The California Supreme Court in 1971 gave Whitney the victory. The patent from the governor, as from the king, came burdened by the public trust. Furthermore, the scope of that trust had grown with time:

> Public trust easements are traditionally defined in terms of navigation, commerce, and fisheries. They have been held to include the right to fish, hunt, bathe, swim, to use for boating and general recreation purposes the navigable waters for anchoring, standing or other purposes . . . the public has the same rights in and to the tidelands.
>
> The public uses to which tidelands are subject are sufficiently flexible to encompass changing public needs. In administering the trust, the state is not burdened with an outmoded classification favoring one mode of utilization over another. . . . There is a growing public recognition that one of the most important public uses of the tidelands—a use encompassed within the tidelands trust—is the preservation of those lands in their natural state, so that they may serve as ecological units for scientific study, as open space, and as environments which provide food and habitat for birds and marine

life, and which favorably affect the scenery and climate of the area.

Whatever the limits the courts have placed on the states and Congress, many owners of dry sand beach property try to extend their property lines, or at least their control, onto the tidelands. The sign declaring PRIVATE BEACH, NO TRESPASSING often stands with its feet in the surf, implying that the upland owner will prosecute anyone passing through on either side of the sign. Owners of commercial piers feel that it is their right and duty to police everything under and around the pier. The notorious owner of a North Carolina pier had for years kept surfers, swimmers, and boaters off his beach and away from his pier with big signs and equally bold threats of violence. The Coke bottle with which he threatened one of the authors who brought a class to look through the parking-lot fence at the pier, had earlier made him the center of a lawsuit. When he saw a young man paddling a surfboard toward his pier he picked up a Coke bottle, took practiced aim, and beaned the paddler on the head. The young man, who was attempting to paddle from Coney Island to Florida, was rescued by a fisherman who jumped into the water. The pier's owner found himself before a judge and jury, being sued for $32,000 and unwillingly reaffirming the public trust doctrine. The North Carolina Supreme Court eventually ruled that owners of piers had no right to obstruct public passage. The court added that piers should be built high enough for people to walk under and for small boats to pass through.

For the most part, the forays of private landowners into the tidelands are limited to seawalls and groins, which the public, no wiser than the property owner, accepts as self-defense and necessary obstacles. The real battle rages back and forth across the dry sand beach over which the public must pass to reach the high-tide mark. And when the tide is up, the public retreats to the dry sand—whether or not it is privately owned.

In many states conventional wisdom says, "You can't stop a man from getting to his property." Whatever the written law, the public, including most landowners, accepts the right of property owners to reach their land even if they have to cross other private lands. The belief seems to apply to individual landholders only. When the public owns land next to an individual's property the damage of providing access often seems unreasonable. For almost two hundred years the public was reluctant to assert its right of access to the tidelands. A few minor legal skirmishes established no clear victory or strategies.

Suddenly in the 1960s and 1970s fighting broke out on all coasts. Scholarly law journals as well as newspapers and magazines published frequent narratives and analyses of the battles. The motive for the public uprising is not complex. In the affluence and population boom following World War II, public use of the beaches leaped dramatically at the same time that developers were closing more and more of the traditional access points. In the realm of beach property the "have nots" enormously overwhelm the "haves"; and since these "have nots" are generally well-fed, well-clothed, well-housed, and educated, they are more aggressive and politically powerful than those people we usually call "have nots." Typical of middle-American activism, the battle for the beaches has been waged by legal swords and lawyer proxies for the combatants. To secure the right of access to the tidelands, the public did not at first forge new laws but picked up the rusty but salvageable swords of the same English common law that had forged the public trust doctrine. The battle still rages, and what was won yesterday may be lost tomorrow.

To date, the deepest wound in private property has been made by the principle of *implied dedication.* Most readers will be aware that local government requires developers to dedicate subdivision streets to public use and maintenance. This, of course, is mandatory and recorded in the county records just as a deed is recorded. Street dedication is an expected cost of doing business. On other occasions landowners give away land or the right of access across land (an easement) voluntarily—sometimes altruistically, sometimes unconsciously. If a landowner allows the public to consistently use a certain access to the beach (or anywhere else), the courts may interpret his action as a donation or implied dedication. In 1970 the California Supreme Court decided two cases (Gion *v.* City of Santa Cruz, and Dietz *v.* King) which have since provided a model for decisions in other states. The city of Santa Cruz had an easy time proving implied dedication of three valuable oceanfront lots. Its garbagemen, lifeguards, and highway crews testified that for at least five years prior to a challenge by Gion, a recent purchaser, the city had maintained a parking area, policed the beach, and provided rock for erosion control. A stream of other witnesses, including former owners of the Gion property, testified that the public had parked on the lots since 1910 and used the beach for bathing, fishing, and picnicking. In deciding the battle between Mr. Gion and the city, the court had no doubt that the rules that applied to dedicating highways also applied to beaches:

> With the increased urbanization of this state, . . . beach areas are now as well defined as roadways. This intensification of land use combined with the clear public policy in favor of encouraging and expanding public access to and use of shoreline areas leads us to the conclusion that the courts of this state must be as receptive to a finding of implied dedication of shoreline areas as they are to a finding of implied dedication of roadways.

In the second case (Dietz *v.* King) the public established its interest without the help of city services or money. North of San Francisco in Mendocino County, where much of the beach is backed by bluffs and magnificent redwood forests, access routes to the beach are few, old, well-established, and more and more frequently used. In 1960 when the public began to annoy the Kings, the new owners of a beachfront property, Mr. King blocked the rough road to Navarro Beach with chains and logs. The roadblock was removed by visitors and reconstructed by King several times until, in 1976, King hired a bulldozer to erase the road and push up a blockade. King claimed that the road had only been opened to the public by the good graces of former landowners and that the public had no established right to the beach. Dietz, suing on behalf of all users of the beach, documented a hundred-year history of public use. It included commercial and sports fishermen, swimmers, campers, driftwood and firewood collectors, divers, and some who came to decorate the crosses of shipwrecked sailors and settlers buried in a small cemetery. Even the Indians had continued to use the road for kelping and fishing as late as 1950. King won his case in Mendocino Superior Court, but was reversed on appeal when a higher court confirmed that lapses in the use of a beach in no way revoke an implied dedication.

In Texas a similar case against a developer who put up fences along the shorefront of Galveston Island (Seaway Co. *v.* Attorney General) brought a host of senior citizens to court to testify for the public. One 82-year-old man testified that not only had he traveled the length of the beach in horse buggies, but that he remembered his father's stories of beach travel. The court declared that since citizens had used the beach from low-water to vegetation line unchallenged, a clear dedication existed, just as it did along many country roads and highways.

The Seaway case also used a broader common-law principle. Implied dedication technically rests on the actions or inactions of a landowner, actions which can be interpreted as a donation of one or more of the rights of property use. The public may also establish rights by its action. If a property is used by anyone against the wishes of the owner and with his or her knowledge, the law politely calls that use or possession "adverse." In the Seaway case the public had not only received an implied dedication from former property owners, but its uses of the beach both above and below the high-tide line were clearly adverse to the owners' interests. Since it had exercised those adverse uses for a long period of time, the public had established a *prescriptive easement*, something like the body of a vine lacing itself through the slats of a fence until it cannot be removed without its self-interest being harmed.

The legal sword of prescriptive easement almost succeeded in cutting down a 176-foot-high observation tower built at a cost of $125,000 on a portion of Daytona Beach, Florida. The tower's owner, Tona-Rama, Inc., said it could develop its beach as it pleased and had obtained a proper building permit. The court recognized Tona-Rama's title but said the public's continued use had established a prescriptive easement and that the tower should be torn down. In 1974 the Florida Supreme Court saved the tower, but only because the tower's 17-foot foundation was not considered a significant intrusion on the public's rights. Nevertheless, three of the seven justices dissented vehemently. Justice Boyd said:

> Surely, when the present owner purchased the land in question, it was common knowledge that the public had, for centuries, used both the wet and dry sand near the ocean for recreational purposes . . .
>
> If this building be permitted to stand, then the owner might well next decide to erect a gargantuan hotel on the property,

> and the adjoining property owners, demanding equal protection of the law, might then begin to construct a series of hotels along the waterfront—similar to the series that now exists along the east side of Collins Avenue in Miami Beach. This would form a concrete wall, effectively cutting off any view of the Atlantic Ocean from the public. A repetition of the concrete wall created by such buildings would be extremely detrimental to the people of this state and to our vital tourism industry.
>
> In my opinion, the trial court and the District Court of Appeal, First District, were correct in ordering the structure removed, for the reason that it encroaches upon the prescriptive rights of the public.

Justice Ervin added:

> This precedent of the Court majority is a regrettable and unfortunate one which will serve to render more uncertain the rights of the general public to enjoy Florida's prescriptive public beach areas which historically they have so long enjoyed. It will encourage, as Justice Boyd so ably points out, further private, commercial intrusions and obstructions upon public domain areas which have been used as such since time immemorial . . .
>
> With Florida's population burgeoning and its recreational needs multiplying by leaps and bounds, the State's courts can ill afford any longer to be profligate with its public areas and allow them to be frittered away upon outmoded pretexts for commercial exploitation.

In 1969, in its aggressive drive to guarantee public beach access, the Oregon Highway Department sharpened a legal sword that cuts a much wider swath than either implied dedication or prescriptive easement. The department was responding to a suit by Cannon Beach motel operators William and Georgianna Hay. The Hays sued to overturn an injunction against fences on their beach. The state's supreme court stunned not only the Hays but all of Oregon's beachfront property owners when it declared that the public had the right to use the dry sand beach everywhere from vegetation line to the sea. The court said that cases of implied dedication and prescriptive easement were too numerous and contradictory to form a good basis for their decision. Also, the court said, these cases apply only to specific tracts of land. Perhaps frustrated by this legal confusion and annoyed at the prospect of endless case-by-case litigation, the court brought forth the common-law doctrine of *customary rights*. Customary rights arose in medieval England when feudal lords granted villagers manorial pivileges and private-property rights long before there was any formal written system of recording or otherwise recognizing property claims. The doctrine of custom grew out of the feeling that a usage which had lasted centuries must have been founded upon a legal right conferred at some immemorial time, so that "the memory of man runneth not to the contrary." The Oregon court said: "The dry sand area in Oregon has been enjoyed by the general public as a recreation adjunct of the wet sand or foreshore area since the beginning of the state's history." Specifically, the court established the dry sand beach as everything between the mean high-water line and an elevation sixteen feet above that line.

All these battles with the ancient weapons of common law are, as we said, ongoing and far from final. Since the courts only interpret law, we can expect their decisions to change as the judges change.

Oregon has buttressed the Thornton *v.* Hay decision with legislation declaring all beaches to be a public highway between low tide and the vegetation line. The Texas legislature has also attempted to take the fate of the beaches out of the endless seesaw battle of the courts. In 1959 a Texas court upset the old notion that the public lands extended to the vegetation line. Sensing a quick killing, private landowners began erecting walls and fences down to the high-tide line, infuriating fishermen, strollers, and beach-buggy drivers. The Texas lawmakers reacted quickly, assembling in a special midsummer session that hammered out the Open Beaches Act. The new law declared that on all the Gulf Coast beaches the public had, through ancient custom and long use, acquired a prescriptive easement to use the beach up to the vegetation line. Legal scholars say this amounts to lawmakers trying to write history without facts, and that the facts of public use will still have to be established in court. The law, however, does tell shorefront owners whose side the state will take in court. The author of the bill agrees that lawmakers simply wrote a sound hindsight into law. Robert Eckhardt, now a Congressman, says:

> The fact that the beach is a sandy beach, immediately indicates that the beach has been useless for anything but matters related to the sea, and that persons using that beach are persons who are there for recreation, fishing, drying nets or various purposes other than, for instance, grazing, which was the littoral landowner's ordinary use. . . . Thus, once you show that the land was a sandy beach you don't have to come in and show a long line of history in which people have used it. It becomes necessary then for the private owner to prove it was not used as most beaches are ordinarily used by people in general.

A 1971 Open Shorelines Act in the Virgin Islands begins with the recognition that "The sea has brought to these islands all of the seven flags that have reigned over them." The intent of the new law was simply "to preserve what has been a tradition and to protect what has become a right of the public." Anywhere in the islands you may use the beaches from the low-tide line to the first line of continuous vegetation or a natural barrier.

In Massachusetts, where a 1614 colonial ordinance established private property limits at the low-tide line, State

Senator William Bulger has been pushing a ten-line bill that has upset most shorefront property owners. Bulger's plain talk states simply: "The public shall have an on-foot right-of-passage along the coastline of the commonwealth below the vegetation line."

What cannot be accomplished by the states could be realized by passage of the federal open beaches bill introduced in several sessions of Congress by Representative Bob Eckhardt, who wrote the Open Beaches Act for Texas. Senator Henry Jackson has sponsored a similar bill in the Senate. Some legal scholars claim the bill is on firmer constitutional ground, since the commerce clause of the Constitution casts Congressional authority over the interstate movement of goods and people across the beaches. Given the fact that 50 per cent of the nation lives within fifty miles of the coast, access to the beach is a "national concern," an epithet that has justified federal jurisdiction in many cases.

Political realists are not waiting for a federal open beaches law. Some people believe so strongly in the public's right to reach the water that they have made beach access a civil rights issue. Two cases in 1972 resounded across suburban shorefront America like a double-barreled shotgun. In Neptune City *v.* Avon-by-the-Sea the New Jersey Supreme Court said local residents (through their town government) could not exclude nonresidents by charging them higher user fees than residents paid. On Long Island a young American Civil Liberties Union attorney, Lawrence Sager, gave the city of Long Beach a legal pummeling for banning nonresidents (Gewirtz *v.* City of Long Beach). Sager made the following points that are sure to be made again in future cases:

1. Long Beach had implied public dedication by a long history of allowing everyone to use the beach.

2. As a creation of the state, a city cannot deny the use of public trust lands to other citizens of the state.
3. The beach had been built up by erosion-control projects and thus rested on tidal lands held by the state for all its people.
4. Barring nonresidents denied them the equal protection of law guaranteed by the Fourteenth Amendment to the Constitution.

Sager has followed this victory with others, and his tactics were used by the New York Civil Liberties Union to open beaches in Rye, and by the Connecticut Civil Liberties Union in Fairfield. New Jersey, with a heavy population within a few miles of its popular beaches, has declared nonresident restrictions "repugnant" to the notion that all its citizens are equal.

In 1971 California citizens, angered by constant defeat of land-use legislation to protect beaches, gathered half a million signatures to place the historic Proposition 20 before the voters. The new law proposed state contol on all building within one thousand yards of the shoreline. Realtors, oil interests, the building trades, and power companies tried to ridicule Proposition 20. The high-pressure ad firm of Whitaker and Baxter co-ordinated several million dollars' worth of billboard, radio, television, and newspaper advertising against the Proposition. This massive ad campaign only helped to reveal to voters the intensity and size of the interests that wanted to control the beaches. Proposition 20 passed by a narrow but firm margin and became the basis for one of the nation's most sophisticated coastal zone management plans.

A similar citizen initiative in Washington grew out of discontent with the state's already lapsed Shoreline Management Act. Although placed on the ballot by 160,000 signatures, the initiative to create a 500-foot control zone failed, partly because alarmed legislators had also placed revisions to the Management Act on the same ballot. The revision, which passed, created a 200-foot control zone. It also required local governments to adopt plans to provide public access, increase recreational opportunities, preserve the natural character of the shoreline, and promote long-term over short-term benefits.

Anyone who has ever cleaned up a public beach knows that public access and use do not always enhance the beauty of a beach. Private landowners have good reason to fear both the numbers and habits of the people who use the public beaches. Most of the nation's largest cities are within an hour's drive of the beach. A quick list includes Boston, Providence, New York, Philadelphia, Baltimore, Washington, Norfolk, Miami, Tampa, Mobile, New Orleans, Houston, Los Angeles, San Francisco, Portland, and Seattle. Even without roads, the public has turned hundreds of miles of open beaches into garbage dumps. Sailboats, speedboats, and rowboats bring so many picnickers and campers to the bay side of Shackleford Island, one of our favorite beaches in North Carolina, that the strand often looks like a landfill. Even the low windswept Core Banks just north of Shackleford was strewn with abandoned beach buggies and dilapidated fishing shacks before the National Park Service took the island as part of Cape Lookout National Seashore in 1978. As the courts strike down some barriers to the public, others are going up.

The implied dedication that rests on actions of landowners may be abandoned by a public that does not use it. In the absence of the public, a landowner may extinguish the dedication by enclosing his land or otherwise signaling intent to exclude the public. So, too, prescriptive easements gained by continued "adverse" public use may be lost if the public abandons the easement or if a landowner (by fences, buildings, seawalls, or fill) establishes a use adverse to the public interest.

Despite recent court decisions, exclusive suburbs and resorts are also striking back. Fashionable and literary Westport, Connecticut, has reversed its usual liberal politics to enact a $25-a-day nonresident parking fee. In Deal, New Jersey, nonresident guests at an oceanfront home found that their use of the beach quickly attracted the local police and a $200 fine. In Greenwich, Connecticut, a yachtsman rowed ashore on a private beach and went to dinner in town. When he tried to cross the dry sand to row back, the local police caught him trespassing and guided him to the nearest public beach so he could swim to his yacht.

The conflict on the beaches is spreading, not abating. Some states and the federal government tied their hopes for law and order to coastal zone management plans—land use plans for coastal areas. Most of the state plans, now in effect or being written in all coastal states, grew from seed money provided by the Coastal Zone Management Act of 1972. Although land-use planning is not politically popular in most states, Congressmen had an unusual tool in preaching coastal planning. The federal government, through its constitutional power to regulate commerce and activities affecting navigable waters, had great but unused authority over most of coastal America. Recent court decisions had given the U. S. Army Corps of Engineers the power to deny dredge and fill permits because of the overall environmental and social consequences of a development. This is a virtual veto power for coastal development, which derives most of its value and success from marinas, canals, harbors, channels, seawalls, and piers. Since the Coastal Zone Management Act embodied a Nixonian federalism allowing states to write their own plans, Congressmen argued that it was a return of federal power to the states. The federal government does demand that state plans address specific questions such as power-plant siting, oil pipelines, wetlands management, and water-pollution control.

The states were slow in developing their own plans, and many were worried that the requirements to lay down guidelines for offshore ports, nuclear plans, refineries, and pipelines would reduce their options in dealing with such development. In 1976 Congress sweetened the invitation with amendments. Included was $1.2 billion in energy impact aid to help states provide public facilities for the energy industry. Equally important, Congress added that planning for beach access and erosion control had to become part of all plans. To back up this provision, it provided $25 million which could be used to buy access if the state put up a similar amount.

Critics of the Coastal Zone Management Act point out that it evades any responsibility for creating a vision of what the shoreline is and what it should be like, but leaves its fate to the fractionated planning of many states. We noted the truth of this argument in our chapter on barrier islands when we pointed out that not one state plan recognizes an island as a unified natural system. This is like trying to nourish human health by writing one diet for the arms and another for the face and a third for the feet.

Coastal zone plans, like most law, considers the beach as several separate pieces of real estate. This reflects the way in which commercial interests appraise and develop the beach. Coastal zone plans contemplate fixed lines and encourages everyone to feel that the uses of coastal property can be spelled out once and for all. This is a fiction that the courts also have indulged in. The fiction does not fit the facts, nor does it help the landowner whose property is on the move to find order or comfort in the maze of case law dealing with the gain and loss of coastal land.

To date, we have not even been able to decide definitively where the high-tide line lies. Is the line drawn at the swollen spring tide, at neap tide (when the difference between high and low is smallest), or at flood tide? Washington state equates mean high tide with the vegetation line. The federal government takes the average height of all tides over a full tide cycle of 18.6 years. Some states say the vegetation line is clearer and shows the most landward reign of the tides. Some lawyers lean toward uniform adoption of the mathematical federal standard, but in many states such a standard would mean redistributing considerable property. A more politically feasible approach would be to clarify present definitions. This too might force some property redistribution, though the legislative debate about tides would educate a significant number of lawmakers about the subtleties of the oceans.

Whether we settle for vegetation line or mean high-tide line, a migrating, shrinking, or growing beach makes any line definition only the doorway to a larger problem. If a beach is eroding or migrating, the tideline or vegetation line will also move. Fixed property lines stay in place. Have the public trust lands and the public easements on the dry sand simply advanced? Or have the public's rights disappeared and the shorefront property owner come closer to the water? Can a property owner build a seawall or bulkhead and fill behind it to maintain his dry land? Or would that be an illegal enclosure of the public trust land?

Erosion or migration poses the most difficult questions. The retreating dunes and vegetation line leave before them a sand beach where privately owned upland once existed. If the public had claims on the former dry sand beach, are its claims good on the new beach? A prescriptive easement based on a period of past use would seem worthless. Implied dedication fails, since probably the landowner has carefully avoided any act which would suggest agreement to public use. Customary rights from time immemorial hardly applies, since the public hasn't ever used the new beach. We have just discounted the three principles which underlie public access and use of most beaches in Texas, California, Florida, and several other states.

Besides erosion the courts recognize three other changes in coastal property:

ACCRETION: gradual additions to land from deposited sediments.

RELICTION: gradual extension of land as water recedes.

AVULSION: a sudden change in the shoreline caused by storms or flooding.

The courts have generally held that avulsion does not change boundaries, at least not if the property owner is careful about maintaining the identity of his claim. A beach growing by accretion or reliction presents some new land whose ownership may be up for grabs. If the new land is privately owned, does the old public easement remain intact, high and dry behind it? This problem becomes especially acute when the new land appears in the shelter of a groin or jetty. In general, the courts have denied private claims on land created artificially, especially by a public works project. Yet the landowner fronting a section of beach which is gradually building seaward because the wind or waves are depositing new sand has little trouble making nature's gift his own.

The law, however, is not clear. To date, no appeals court has given a definitive ruling on the fate of public rights on a migrating beach. Nor have the public lands been defended successfully against seawalls and other structures that protect private land while encouraging the waves to carry off public lands. The most promising idea so far is the "rolling or floating easement." Supporters explain that because of the shifting nature of sand, the public never visits the same beach twice. Their easements established by use and custom are moving easements on the beach, wherever it is. This would clearly favor the public where the beach is retreating. It would just as clearly favor the property owner where the beach is growing. Given an almost universal retreat as sea level rises, the rolling easement is in the public's favor.

Here we come to the ugly duckling of coastal planning. If coastal management efforts recognize the dynamic nature of the beaches, then planners and politicians have to abandon the pretense of neutrality. They must decide whether changes in the shape of the beach, especially erosion, are changes in public rights or private plots.

To decide in favor of the landowner could soon leave us with many people holding deeds to pieces of the sea itself, titles hard to defend or enjoy, no less stake out. Since the beaches are in fact already ruled by the sea, the best guide to public policy may be found in the work of a seventeenth century scholar from the Dutch seaport of Delft. In the midst of a growing international fight over the rights to shipping lanes, Hugo Grotius argued that the oceans could not be legally possessed by mankind. In *Mare Liberum* he refuted Portuguese claims to specific areas of the ocean, saying:

> . . . that which cannot be occupied, or which has never been occupied, cannot be the property of any one, because all property has arisen from occupation . . . that all that which has been so constituted by nature, that although serving some one person, it still suffices for the common use of all other persons, is today and ought in perpetuity to remain in the same condition as when it was first created by nature.

And again, in *Rights of War and Peace*, he said:

> There is a natural reason also, which renders the sea, considered in the view already taken, incapable of being made property: because occupancy can never subsist, but in things that can be confined to certain permanent bounds.

The logic which Grotius exempts oceans from the realm of property can also clarify our thinking about beaches. Though the beaches have been occupied, the occupation is seldom permanent. It endures longer than the floating of most boats on the open sea, but in many places it does not last long enough for even one generation to pass its possessions on to the next. An occupation of the beaches is a state of war between mankind and nature and between human beings. And however successfully people occupy the beach, they have never been successful in confining it to "certain permanent bounds." Because we cannot confine it or peacefully distribute its shifting surfaces, we might best agree now that it is owned by none of us and by all of us.

12

How to Live with a Beach

IF OIL RIGS, deep-sea ports, and atomic power plants can stand in the ocean itself, can't we, with better design, build houses, hotels, and motels that can stand on the beaches against wind and surf? Yes, it can be done. Super buildings and super defense structures could be built on foundations anchored below the moving sands. In fact, the failure of each engineering project has led to bigger and more expensive projects. The end of this trail is the invulnerable fortress. Engineers, given unlimited resources, have the sophistication and skill to create development that can survive by more than chance. But such super building, to pay for its security, would have to generate the super income of an oilwell, or be backed by the guaranteed return of a public utility. Such possibilities, like sand castles, are subjects for fantasy.

Naturally, coastal development cannot afford perfect safety. The costs of perfection in any technology are ruinous. But the two crucial economic questions for coastal policy are: Against how much risk can the developers afford to protect their clients? and, Should the public pay for any part of this protection or its consequences? For both developers and homeowners, the amount of affordable risk is simply a question of adding safety items onto the basic building budget until the expected return no longer justifies the expected benefits—whether those benefits be the sound and sight of surf beyond the living-room window or the profits from free-spending tourists.

A recent article in *Geology* by Dr. Christopher Mathewson and Donald Piper, both geologists at Texas A and M University, proposes a system for measuring risk against how many dollars development could generate. Using Padre Island, Texas, as an example, they measured the degree of risk in each of eight beach zones and also the cost of building in that zone. Costs included type of structure and access. High cost and high risk meant low development value. Not surprisingly, the dunes and beaches had a low development value. They concluded that by building reinforced-concrete high-rise buildings on the foredunes and beaches, developers could generate a good income and escape most disasters. Seawalls and bulkheads would be an economically small part of the financial picture. In fact, Piper and Mathewson say such high density makes it "economic to stabilize or remove the entire dune field." They also point out that dune removal makes access less expensive.

Three geologists from the University of Texas in Austin thought Mathewson and Piper were irresponsible. They wrote to *Geology* and accused their colleagues of "simple cash-flash real estate economics," encouraging developers to build disasters. The challengers said the authors ignored the dynamic nature of the beaches. The Mathewson and Piper approach grows out of the fatalistic notion that development can't be stopped and that all benefits in life derive from dollars. Dr. Mathewson has written elsewhere:

"We have all heard the statement 'It isn't nice to fool with Mother Nature,' but fool we must if we are to develop the fullest potential of our coastal lands."

Mathewson has not always been so encouraging to developers. In 1975 he told a seminar of coastal experts that only two structures could survive on the beach—a tent that bends with the storm, or a concrete monolith. The first he considered impractical, and the monolith "economically impossible." His middle ground, contrary to his Austin critics, does not ignore the inevitable moving of the beaches. It only says that if a project generates income fast enough, investors have a good chance of recovering costs and a nice profit before the sea gains its inevitable victory. A number of developers selling expensive condominiums and motel operators renting rooms at $80 a day have told us that if they survive five years, let the ocean come. By that time they will have recovered their cash and banked a profit. The portion of their costs financed by loans would be covered by insurance.

Mathewson and Piper have provided a useful process for answering our question about how much protection developers can afford. Their conclusions, however, are economically naïve and politically irrelevant, because they do not consider the cost of government subsidies to development. Nor do they talk about the costly damage development does to beaches around it or to the public trust lands.

The greatest subsidy to beach development has been the innumerable projects built by the U. S. Army Corps of Engineers. The Corps itself, in its huge 1970 *National Shoreline Study*, recognized the basic truths of the dynamic equilibrium, but its conclusions were traditional and offered little change from the strategy prevalent when Trajan first tried to save the silted-up port of Ostia in the first century A.D. At the same time, the Corps has begun to publish booklets describing what it considered a new environmental mission. Two conflicting policies now stand side by side. The Corps by its actions carries on its old romance with "stabilization" while hesitantly pledging faithfulness to a new bride.

Local Corps offices live with all the daily pressures of threatened beach communities and despairing property owners. Although in theory and law the Corps never initiates beach projects, it often seeks to enhance its image and support its engineers by suggesting to the local chamber of commerce, the town board and civic clubs beach-protection measures they might request. The Corps is also usually unwilling to take a comprehensive view of new development. Through its permit authority it holds absolute power over all environmental effects of development, and its regulations state that the Corps must consider not simply the request for a marina or pier but all the consequences of the development using the facility. With very few exceptions, the Corps has preferred not to look at ravaged maritime forests, bulldozed dunes, dense development, and water pollution that surround the piers, canals, marinas, and channels that beg its permits.

Once established, most developments, as well as older communities are covered by government-backed insurance. The National Flood Insurance Program (NFIP) has recently improved its standards for insurable buildings, but this does not discourage development, nor is that its intention. The Flood Insurance Program does not prohibit building in dangerous beach zones. It simply requires stronger construction. To this extent it increases survival chances in minor crises. The more costly buildings also mean development, and government mortgage and insurance subsidies are more than ever a privilege of few Americans. In many cases the Flood Insurance Program is a paper seawall. A local building inspector in Hancock County, Mississippi, told a colleague of ours he had never heard of the hurricane building code in his county, a code which is famous among planners for its tough standards. Most officials of the NFIP would like to deny insurance to communities which do not take strong action to actually prohibit development in danger zones. Congress, however, has been pressured into denying such power to NFIP. New standards do require that local government must prohibit the destruction of mangrove stands and sand dunes in order to receive flood insurance, and regulations have been rewritten to encourage better building practices and higher premiums in high-risk areas. When disaster strikes, the higher premiums will save the government, but they will save neither the beaches nor the buildings.

Other government subsidies include EPA grants for sewage plants which serve development, maintenance of the Intercoastal Waterway, Coast Guard bridge permits, and Farmers Home Administration grants, and low-cost loans for streets and water supplies.

If anyone were taking bets on whether man or the sea will win the struggle of the beaches, we would bet heavily on the sea. Rising sea level, even within a single generation, will drive back many beaches, undermine cliffs and bluffs, and flood the mangrove swamps. Its effects will destroy or render impotent artificial dunes, seawalls, jetties, and groins. As we write this final chapter, our bet is made all the more certain by the unusual absence of major coastal storms for almost ten years. (Hurricane Ella, recently stalled off the Carolinas for two days, at the last moment turned northeast, sparing the mid-Atlantic coast by a hair.) How to live with the beach is embarrassingly simple: allow the dynamic natural system to re-establish itself. We are not proposing a hopelessly romantic solution. Behind our conclusions lie all the detail of the preceding chapters, all the "war stories" we have told and hundreds we have not told, all of the scientific books and periodicals we have read over the years. We are, of course, also reacting to the almost complete failure of engineering attempts to manage the beaches, a failure which has left more coastal communities in greater danger than ever before. And with less sandy beach. We are not alone in our conclusions.

When coastal zone management laws first began to surface, many experts feared they would simply enlarge past failures, or become meaningless red tape. In May 1972, one hundred scientists met at the Woods Hole Oceanographic Institute on Cape Cod to consider what principles should guide the new technology of coastal management. The scientists concluded that the best idea was to preserve or imitate the natural system as closely as possible.

Proof, of course, lies in application. This prospect troubles a country that has placed so many houses and businesses where nature will not tolerate them. Our approach, nevertheless, is already getting a limited but nationwide trial in natural laboratories on the Atlantic, Gulf, and Pacific agency with budget troubles.

Before 1952, the National Park Service had developed its policies entirely through its experience with inland parks and monuments. In Yellowstone and Yosemite stability meant preservation. Naturally, when the Park Service began acquiring Cape Hatteras National Seashore, it tried to maintain the artificial dune wall built in the Depression. But preserving "stability" in this seashore and nine others that followed began to eat a big hole in the federal purse. The U. S. Army Corps of Engineers was proposing another fifty miles of dune dikes for Cape Lookout just south of Hatteras, and another dune line along all of Fire Island,

New York. Assateague Island in many places already had a barrier dune wall like Hatteras.

Park Service officials in Washington began wondering if they hadn't been saddled with the job of managing disappearing land. Meanwhile on the beach, scientists in and out of the Park Service were studying sand movement and cores taken out of the islands. They watched vegetation grow up through two and three feet of overwash sand. They were poring over rolls of old maps and peering through microscopes at little fragments of gray shell and minute black specks found in peat clumps. In many monographs and articles with esoteric titles they talked to each other about pollen spores and layers of sand and the rate of grass growth. Many politicians and "hardheaded" business people might have seen in all this just egg-headed boondoggling. But this research produced the raw material for the story of how barrier islands move. In 1973 Dr. Robert Dolan, a young scientist at the University of Virginia, distributed a paper among Park Service officials telling them what the new beach story meant for national seashore management. Acting with unusual speed, Park Service Director Ronald Walker issued a new management policy for Cape Hatteras National Seashore. Dolan's paper had a dull title, "A Strategy for Management of Marine and Lake Systems Within the National Park Service," but the new policy almost started a war.

Local residents, developers, and business people quickly saw that if nature took its course, as the new policy proposed, the ocean would soon be at their doors. If the new policy works well, the famous Cape Hatteras Lighthouse will fall into the Atlantic Ocean before 1985 although the Park Service is also considering moving the tower. Outlying parts of the towns of Buxton, Avon, Nags Head, and Kitty Hawk will also disappear. During storms, waves will break through the artificial dune dike and overwash parts of the islands, including the roads which link them to the mainland. But if the new policy works, the long narrow island will be healthier in 1985 and ultimately more stable. Fewer tax dollars will be spent rebuilding dunes and beaches. Moreover, few lives will be in danger.

As natural processes re-establish themselves, the dominant force in the environment, controlling both the human and natural environment, will be the ocean. As the high narrow dune dike artificially built during the Depression breaks down, it will form a series of smaller irregular dunes along the ocean. Because these new foredunes will not be completely covered with fertilized grasses, sand will blow over them from the beach. And yet more sand will wash across in rough weather as the waves top the dunes. A system of secondary dunes will help sap the energy of the advancing sea. Where sand does wash completely across the island it will spread out in the sounds. Grass will then grow and new marshes will form, and eventually new land.

While the natural environment and the public lands might be saved by the new policy, some permanent residents and businessmen see themselves as live sacrifices to a bureaucracy mesmerized by radical environmentalists. The older parts of villages on the seashore were prudently built far back from the ocean. For at least a hundred years they will survive all but the greatest storms with only wind damage and minor flooding, but motels, hotels, stores, and beach cottages built close to the back of the dune ridge will certainly disappear. As Dr. Paul Godfrey points out, their builders were just as mindless of the sea on the other side of the dunes as were the terrestrial plants which began to grow much closer to the ocean than normal because of the temporary protection of the dune wall.

When director Ronald Walker first announced the new

policy in 1973, it so shocked traditional notions of how man should deal with nature and how government should protect its citizens that it heated local tempers to the boiling point. Some Park Service administrators said they were afraid to park their officially marked vehicles in front of the restaurants where they ate. Things will probably get worse when the full implications of the new policy sink in. To date, local people still feel that the federal government cannot completely abandon its efforts to fortify the beaches. Congressmen, reacting to local emotion and oversimplified reports of federal "retreat" on the beaches, have promised to save everyone and turn the Park Service around. And state officials continue to reassure everyone that the developed areas of the island will not be abandoned, though they are deliberately vague about how the state can protect development. Where it could find the money is also a big question.

Between 1957 and 1968 the Park Service was spending some $420,000 a year on its seventy-four miles of Hatteras dunes and beaches—a relatively small amount. From 1968 the costs rose sharply as the beaches continued to disappear in front of the dunes and the dunes themselves were undermined. In 1971 the Corps spent a half-million dollars to repair just a few thousand feet of eroding shoreline near Buxton and Cape Hatteras Lighthouse, and the dunes were being supplemented with groins. The groins and the emergency repairs both failed. The $4.3 million price tag on the final 1973 beach replenishment project in that same area should in itself be a warning that neither federal nor state government can afford to keep battling the ocean—even if they could win. The $4.3 million would have been more than enough to buy up most of the buildings damaged by the 1973 winter storms in the Buxton area. In 1975 all the private motels in Buxton were assessed at less than $320,000.

However gently the Park Service states its new policy, in reality it says this: "We have no obligation to save private property, and if the beaches move out from under your home or business, or from under our lighthouse, we will only stand by and watch them fall, because in the end this is the most economical kind of management, the most beneficial for the natural environment which brought all of us here, and the quickest way to forestall future development and ever greater danger to property and life."

Park managers at Assateague Island, straddling the Maryland Virginia line, took the new policy so literally they began bulldozing artificial barrier dunes. Park Service scientists and public outcry soon stopped the work. The transition to a natural system cannot be made so abruptly, and the Park Service says that in some places the alterations are so great, and the public interest in a status quo so strong, that return to a natural system is impossible.

The Fire Island National Seashore has suffered severe erosion ever since jetties were built to stabilize Shinnecock and Moriches inlets. Affluent, sophisticated, and often eloquent New Yorkers have crowded the thin, delicate island with a small city of cottages and villas, disrupting the island's vegetation, drainage, and sand movement. In 1977 the Park Service came up with a plan that provides what might be called "planned natural protection." Sand stopped by the jetties will be pumped back into the longshore currents. At the same time, along most of the island the Park Service will no longer build up eroding beaches, pumping sand from the ocean floor. New inlets that open during storms will not be automatically closed, but steadied to determine if their role is constructive. To avoid one section of beach being built at another's expense, no groins, bulkheads, revetments, or seawalls will be allowed. Dunes get special protection with limited hiking trails and public crossings. Specially protected dune districts guard areas most subject to storm damage. To protect the marshes which build the island's girth, new mosquito ditches are forbidden and the old ones will be allowed to fill up. Both herbicide and pesticide spraying have been stopped. The most innovative part of the plan is a proposal to build new marshes. Sand dredged from inlets and boat channels would be spread like artificial overwash fans on the back side of the island. The new marshes would presumably widen the island, prevent breaching, and eventually provide a foundation for new land as the island retreats.

The cautious simplicity of the Park Service policy stands in blunt contrast to the erosion-control plan offered for the same area in 1978. The Corps, responding to pleas to save eroding beaches and property both on the Fire Island Seashore and farther east, has proposed a massive "stabilization" program which includes up to fifty groins and elevating dunes in Fire Island Seashore's natural area to twenty feet. The Corps estimates that its work will initially cost $139 million. Every year pumping sand onto the beaches will cost an additional $8.5 million, with local government shouldering all of these costs after ten years.

At the same time the Corps proposes to cure shoaling in Moriches Inlet by extending its jetties another 400 feet at a cost of $4 million. The extended jetties would trap even more sand that might reach Fire Island.

The Corps' proposals rest heavily on its contention that benefits of its projects always exceed costs. More and more state and local governments are questioning the Corps' economics.

In 1976 when the Corps proposed a protection plan for Delaware's shoreline, scientists, planners, and engineers at the University of Delaware and in state government discovered gaping holes in the economic benefits attributed to dune building, bulkheading, and beach nourishment. The Corps had estimated construction cost at $36.1 million with no room for cost overruns. The interest assumed for the state's fifty-year construction bonds was 3.25 per cent. The Corps estimated state maintenance costs for the first year at $2.1 million, and property value saved at $2.7 million. The director of Delaware's planning office pointed out that every year the state would only be paying $2.1 million in maintenance (assuming no inflation) but over a million dollars interest on its bonds. Thus, to prevent annual property damage of $2.7 million, the state would spend $3.1 million. The Corps' estimates also assumed that after any storm, houses, hotels, and businesses would be rebuilt as they first existed so that the cost of coastal catastrophes would always be the same, unreduced by either building codes, stricter insurance regulations, or any application of common sense. Nor did the Corps consider the likelihood that a major hurricane could completely undo all beach nourishment and much of the dune building and bulkheading. The state's final analysis was that the whole scheme would cost over half a billion dollars as opposed to the Corps' estimate of $140 million.

The contrast in policies between the National Parks Service and the Corps is not simply between two agencies, one with a bias toward engineering and the other with a back-to-nature approach. The Park Service has recognized that there are places where natural maintenance cannot cope with human needs or where it is simply inappropriate.

In its management of the Gateway National Recreation Area surrounding New York Harbor, the Park Service has had to deal with irreparable changes in the natural system. At Floyd Bennett military airfield opposite Kennedy Airport on Jamaica Bay, the Park Service at first intended to tear down the concrete seawalls and bulkheads. The natural marsh fringe, they reasoned, would re-establish itself with all the benefits of a marine life nursery. Fortunately, scientists looking into the bay's natural history discovered that the field had been built entirely of artificial fill. Exposed to the swift currents of Jamaica Bay's inlet system, most of the airfield would have eroded away long before new marshes could have grown.

So far the Park Service has avoided the bane of all nervous small bureaucracies—falling in love with the substance of its successes. It not only recognizes the differences in man's influence on the beaches, but recognizes that no two natural beaches are alike. A policy that prescribes the same management for all beaches, pristine or developed, north or south, cannot be in tune with the varied intentions of nature. Whereas the Park Service will no longer stabilize dunes on the Outer Banks, the goal on Cape Cod is to reforest and arrest a large area of moving dunes. The job is a formidable undoing of 350 years of New England's history.

For several years botanist Cheryl McCaffrey, under the direction of Dr. Paul Godfrey, studied both the human history and the soil profiles of Cape Cod. Her studies indicate that the natural condition of the Cape does not include the huge naked dunes which now rove the north end of the Cape near Provincetown. Her soil cores confirm that beach and maple forests existed for thousands of years where now there is only bare sand. And the historical record not only shows that mankind has stripped the land, but that we have ignored the best advice of experts since the founding of the Plymouth colony.

In 1654 the Cape became the New World's first natural preserve when colonists purchased it from the Indians and forbade use of the lands without written permission. Poachers, squatters, and illegal sawmillers mocked the conservation ordinance. Overgrazing and lumbering denuded the sandy hills, and the dunes began to move. In the mid-1700s new laws prohibited grazing but failed to protect the woodlands, perhaps because the hills already looked more like dunes than forests. By 1802, officials published a report describing 1,700 acres of nude sandhills. In 1825 a committee recommended replanting the sand hills with brush fences and grass. During the latter half of the 1800s the state legislature tried to control grazing, woodcutting, and cart traffic—all with little success. Today's tourists, admiring the dunes, seldom realize the scene is an ecological disaster. But the Massachusetts Highway Department which removes 10,000 cubic yards of drifting sand a year from one highway in the area, knows that the north end of the Cape is becoming more and more unstable.

Today the Park Service is trying to remake the northern Cape into something like the forested spit behind which the *Mayflower* dropped anchor. If they succeed, they will achieve the same goal they seek at Hatteras: nature will maintain itself, with benefits for both mankind and the environment.

In the management of the National Seashores we are seeing a rare instance of new ideas originating and proving themselves within the great machinery of federal government. The Park Service is not a developer who must generate profit or a town council elected by property owners. The political pressures and public protest that have rained upon the Park Service gather in even greater clouds over state and local efforts to return the defense of the beaches to nature.

Francis Shepard, the father of geological oceanography, has long lived in a house set back a respectful distance from the cliffs of La Jolla, California. Now in his eighties, Shepard has written about the entire U.S. coastline, and for more than twenty years he has watched the cliffs near his home. He has watched the ocean cut back the cliffs and accepts this inevitable process. Nevertheless, the spectacular view that inspired Shepard's house is so coveted that a developer recently bought the clifftop between him and the ocean and sold it in small lots for prices from $300,000 to $500,000 each. Why hasn't local government stopped the cliff crowders? Restricting development with this kind of money behind it invites sophisticated political counterattacks and lawsuits. To date, few local governments are willing to brave the storm. Nevertheless, we have entered a new era of populism and public thrift, when budget problems and high taxes have begun to help state and local governments, like the Park Service, discover that the natural system, tried and proven over thousands of years, offers a defensible investment with both security and a low price tag.

Just north of La Jolla in the town of Encinitas, where several city blocks have fallen off the cliffs, the board of supervisors recently considered three strategies for defending its ocean frontier. Protecting property by shoring up the bluffs and raising seawalls was dismissed as too expensive and ultimately futile. A *laissez-faire* policy of issuing building permits and simply removing property from the tax books when it vanishes, was set aside as irresponsible. The town did replace the old 25-foot-bluff setback line with 40 feet and declared it would no longer spend money on erosion-control structures to sustain development. On all deeds to threatened property will henceforth be stamped in capital letters: THIS PROPERTY SUBJECT TO LOSS DUE TO CLIFF EROSION. The county's public works people have been directed to take immediate steps to remove houses that have begun to collapse or that sit on lots where the bluff is visibly slipping. To consolidate public support, the planning department is now gathering scientific and historical data to educate the public about the nature of the coast on which they live.

In Florida beach-development pressure is so great that at least one community had to "secede" from its county to protect itself. No sooner had residents announced their intention to vote in a new charter than developers and some public officials plotted to kill the infant city. The bridge from the mainland of Lee County to Sanibel Island was finished in 1963, but until the 1970s visiting shell collectors were the main users. The Gulf Coast development boom, however, was rolling steadily south from St. Petersburg and Sarasota.

In 1973 and 1974 a building boom swept across the new bridge from the mainland onto the six thousand acres of Sanibel Island. Some 2,400 residents watched as subdivisions and motels voraciously began to eat up the thin strip of land between the Gulf of Mexico and the mangrove swamps protected as the J. N. Darling Wildlife Refuge. The county's planning department envisioned as many as 35,000 homes, apartments, and condominiums on the island. Environmentalists joined residents in advocating a new city. The unusual alignment occurred because this was one of those rare instances of a community declaring itself an independent city in order to stop its own growth. On Election Day in 1974, Sanibel's citizens voted to "secede" from the county in order to plan their own future.

The successful birth pays at least partial tribute to the city's most famous past resident, Rachel Carson, author of *The Sea Around Us* and *Silent Spring*. The new town council immediately declared a moratorium on all building until they could draw their own plans for Sanibel. But many people on the mainland, and some islanders holding their land for speculative profit, saw the new city as a dangerous babe in the manger and issued orders for its death. A developer and his lawyers quickly went to court and asked that the city of Sanibel be declared illegal. The Bank of the Islands, whose president was solidly behind the new city, was stopped from providing a $250,000 loan to cover operating expenses until the city could collect taxes in 1975. The new city was saved from withering away by a citizen effort that grows ever more remarkable in these days of anti-government tax revolt. Mayor Porter Goss, a Yale graduate and former CIA employee, called on residents to loan the town the money it needed. Within three days, citizens eager to spite their mainland enemies loaned the city $250,000 and Goss was saying, "No, thank you," to willing lenders.

The city's first big investment was a $106,000 land-use plan drawn up by the famous Philadelphia planning firm of Wallace, McHarg, Roberts, and Todd. The plan considered environmental factors, from offshore sand supply and wave energy to mosquito control in the marshes and hurricane evacuation. After these studies, the Sanibel Planning Commission, with its more intimate knowledge of both the physical and political environment, modified the firm's recommendations. The final blueprint for the future, adopted in 1976, allowed a maximum of 7,800 dwelling units of all kinds, including existing homes. A complex plan of allocating development priorities allowed five times as many homes in the "prime developable" zones as in "hazardous zones." Homes in the hazard zones must have no environmental impact and be built according to very strict standards. Recognizing that beach development destroys the beach, and citing Miami Beach as an example, the plan severely restricts waterfront building, prohibiting it altogether within 100 to 250 feet of the high-water line. As a result, Sanibel's 900 tourist units have a typical winter occupancy rate of around 90 per cent. Florida's highest. The hotel owners prosper accordingly. The planners also recognized that the tax rates necessary to service and protect unrestricted development might tax the city's retirees and poor white and black residents out of their homes.

Property owners who felt the value of their property had been taken away by the growth limits sued under the "just compensation" clause of the Fifth Amendment to the Constitution. Developers have sued to protect their vested interest, but to date the city has won all cases. In 1978 the city turned the tables and sued the Atrium Condominium Complex for building a pool and several decks too close to the beach. The suit cast a cloud over all the titles to the $150,000 condominiums, and the owners quickly settled out of court, agreeing to move the pool and pay the city $62,250. At least one city councilman told us the city felt sure it could have made Atrium tear down whole buildings. Beach preservation with such muscle is so popular with town citizens that when the city's first planning commissioner, Duane White, ran for a vacant council seat, he won with 75 per cent of the vote.

If Sanibel survives the continued legal attacks, it will be the first community to recognize that the beaches and marshes of a coastal island should be preserved in their natural state in the name of both economics and safety.

But even Sanibel, a well-planned town, is not safe in hurricanes. Its highest land, ten feet, hardly matches the possible twenty feet of a major storm surge. Given such inevitable risks, some engineers and geologists have suggested that we simply calculate how to make money at these odds and move ahead with development anywhere investors want to back it.

The local approach to beach protection, if pursued objectively, may be the best way to recognize and cope with the infinite variety of beaches that issue from the constant shifting of the four parts of the dynamic equilibrium. The beaches at the bottom of the cliffs at Encinatas reach equilibrium through a process that is not only different from Sanibel's but even from the beaches a few miles away. But solutions must not only be local, they must contain a sense of history. The character of a beach is not fixed in time like a painting, a sculpture, or a photograph. The character of the beach with its time dimension is a process, not a position. Each beach holds its dynamic equilibrium in its own way, by developing a pattern of behavior that covers both space and time.

Living with a beach, like living with a person, depends upon our knowledge of a changeable character. But most of us are newcomers to the beach, and most beaches we come to no longer show their natural character. Long ago bulkheads, groins, seawalls, and jetties interfered with the natural equilibrium. True wilderness beaches are rare, most of them in Alaska and in the rugged Northwest. A few miles remain untouched at the tip of Florida and along the granite islands of Maine. Yet, despite the changes man has made, it is still possible to know how any given beach, if left to itself, would reach equilibrium.

A well-established principle of geology states that the natural laws now shaping the earth's crust, including the beaches, have shaped the past and will shape the future. Recorded in history, in geologists' files, and in the sea and sands themselves, is the information we need to see through the mask of development and engineering and rediscover the natural beach. Although most of our coastline remained wild until the era of the automobile, centuries of military strategy, shipping, fishing, and shellfishing have yielded a great archive of maps, charts, and photographs. What does not appear on paper is often recorded indelibly in the minds of local fishermen or other natives who must know their environment intimately to earn a living.

So far these records have been gathered and studied mainly by geologists, botanists, and a few engineers. The field work is sufficient to support action. The facts about erosion rates, overwash locations, future inlet sites, inlet migration paths, and flood-prone areas are available to planners, politicians, and the public. These facts should now become the foundation for building and revising the coastal management plans which will soon govern the fate of all the nation's beaches. No better basis exists for any public action, whether it is school curriculum or foreign policy.

Plans based on such information will, of course, show that vast areas of beach are unfit for present and future development. But it will also show that there are areas where development is safe and sane, at least for more than one generation and in some cases for many generations. While good plans will threaten profits made at the public's expense or by destroying the natural environment, in safe areas the plans will guarantee enduring profit. These havens will be high, broad sections of islands and wooded hills some distance behind the beach and dunes; the edges of the bluffs and cliffs will be left to grasses and trees.

But how do we manage heavily developed beaches? What should be done on Miami Beach, St. Petersburg, Ocean City, Virginia Beach, Malibu? Must Cape May, the first beach resort in the nation, spin out its days as a decaying town on the public dole? Must it forevermore peek over the top of a massive seawall, the ocean vista which helped write its glittering history, now threatening to keep the town on the edge of crisis?

For most of these beaches no alternative exists except to return to the natural beach. As sea level rises and beaches retreat, seawalls become the only sure method of protecting development. When the community is rich enough, a beach strand can be maintained by lavish expenditures for pumping up new sand from the ocean. Where nourishment stops, the seawalled beach city quickly becomes a fortress surrounded by an angry sea.

Not that Atlantic Ciy or Miami Beach should be demolished, nor should we refuse to pump another grain of sand onto Coney Island's beach, which is used by as many as 100,000 people a day. But we must let the ocean take over great stretches of developed shoreline, including the motels and hotels. This is the most bitter pill of all to many people. Most of us have a built-in horror of seeing any large building fall down. From this fear comes great, if often begrudging, political support for all kinds of defensive measures to save beach development, especially that development identified with the wealth and "good life" so many people covet. Nevertheless, if we are ready to stand back and let the chips fall, we will find ourselves with opportunities we never thought possible.

At the beach nature sometimes gives us second and third chances. Once development has established itself inland, we are seldom offered a cleaned slate on which to start again. But on the coast the slate is often wiped clean. In New Jersey, on Long Island, and in Rhode Island the 1938 hurricane cleaned miles and miles of heavily developed beaches. Beach development in the Florida Keys has been almost destroyed twice in the last forty years. Hurricanes Camille and Betsy, in 1964 and 1969, both cleared Mississippi's waterfront of thousands of houses. On the West Coast the houses crowded wall to wall at the foot of the Malibu cliffs are all recent successions to earlier cottages destroyed by mudslides and storm waves. Innumerable other beaches will be wiped clean either by major storms or shortly after when stubborn taxpayers refuse to provide emergency sand-pumping and seawalling.

Economics is the basic question, since little waterfront development can claim the right to exist on the grounds of beauty. If we see engineering costs in the context of continued inflation and dwindling supplies of nearby sand, the cost of defending depreciating buildings is seldom returned to the economy through taxes or growth. The return to a natural beach does not mean that all development disappears, but only those constructions whose defense destroys the beach or costs more than the buildings are worth or threatens the rights of future generations. No government body will condemn waterfront property. We simply have to wait for erosion or the next storm to take its toll. Meanwhile, with a little grace, there are some things we can do to ease the pain.

Where the back side of a barrier island or spit is not seawalled, marsh planting and the spreading of sand to imitate overwash can create new land, just as natural migration does. At least some development lost on the oceanfront can be relocated to new land. The money saved by not building new defenses could be used to clean up the rubble of old seawalls and groins. New Jersey's beaches, often littered by the stubs and splinters and rusting slices of several generations of old defenses, would be improved more by their removal than by new construction.

After most disasters we have come back to rebuild. It is the proud, manly thing to do. It is often what the banks require us to do with our insurance money. We must lay aside our pride and stop setting up history to repeat itself. The next time nature offers us a clean or even a partially cleaned slate, let us be prepared to write on it a new idea, one that proves our ability to learn and to live within our means.

We do not expect many property owners to share this desire.

A few years ago when we began making notes for this book, motel owners and subdividers on the Outer Banks of North Carolina threatened to sue one of its authors for stating specifically that certain motels, houses, and lots would disappear in a hurricane. The businessmen agreed that there are danger spots on the beaches, but they denied that their development was threatened. They were understandably outraged at having someone stand on the tidelands in front of a new motel and a luxury subdivision and show a group of television newsmen and journalists how and where inlets would form and where foundations fail. No lawsuits were filed, but the men who threatened to sue reacted to the description of their property as many people will react to good coastal management plans. If a law limits property rights on the grounds that development is dangerous, the law will be quickly called a false prophet and taken to court.

Conventionally oceans, beaches, and natural hazards are seen as fickle and unpredictable. This attitude leads people to say, *Nobody can make predictions.* That argument extends itself to say, *Therefore, no scientist and no land-use plan can say my development is dangerous. It's just a matter of guess and opinion.* All predictions, even tomorrow's sunrise, are matters of opinion, since men and women are not gifted with perfect foreknowledge and nothing is known absolutely until it happens. However, the legal problem faced by coastal management plans will not be to prove they are based on infallible forecasts of the future, but to prove that they are based on an intelligent reading of past and present natural processes. That reading does not leave much room for our present forms of development.

Homo sapiens is a terrestrial animal, and the agents of its evolution have been rocks, trees, and grasses; mountains, prairies, jungles, and deserts. While some peoples have lived on stilt houses above quiet swamps or shallow lake waters, none have ever made a permanent home on the beaches. Nevertheless, today the beaches are more important to us than ever before—not only for the material benefits of flora, fauna, and safe harbors, but also for their power to rest and pacify the mind. To survive in the new environment and to preserve its uniqueness for the great majority who will always be visitors, nothing less is required than that we overcome our natural and cultural heritage.

There is a national park in Key Largo, Florida, that is underwater. There are no motels, parking lots, refreshment stands, or souvenir kiosks. The visitor must adapt to the park's environment: strip down and make like a fish with fins, scuba gear, and masks. Such total adaptation is not required on the beach, but we would do well by ourselves and our environment to shape our presence to the demands of the environment.

As we begin a new kind of planning for human presence on the beaches, that planning must be built upon the fundamental truths we have learned so far. In our imagination we can apply these plans to the beaches of the future.

Where to build. A new kind of setback line will recognize that the zone of highest energy is on the berm and in the foredune area. Here there must be an inviolable setback line, a line that represents the foremost limit of all development of any kind. Setbacks, measured from the waterline, have been tried in several places but usually in an unsophisticated form. What protects a building is not its distance from the sea but the volume and mass of the sand between it and the sea. A thorough study of storm history and the events during normal and large storms will reveal just how much sand is needed between buildings and the sea. A typical setback regulation might read: "No building will be built in such a manner that the dune separating it from the beach shall be less than fifteen feet high and fifty feet wide at the base"—if such measures suited the circumstances.

Buildings. The most ideal form of beach residence is a tent. It occupies little space, can be quickly moved, and its low profile presents little obstacle to sand being moved by the wind. Since the environmental ideal is seldom the public preference or practical, we have to consider at least some means of providing more permanent structures without losing all the advantages of the tent.

Although suitable buildings may vary from beach to beach, large hotels, motels, and condominiums are appropriate only in the highest, most sheltered areas far back from the waterline. This is where early nineteenth-century builders built them. Anything built in low areas, near the dunes, or in overwash areas, are in danger. Numerous and famous "hurricane parties" have proven the fatal results of trying to ride out storms. Similarly, endless stories of vanished houses, parking lots, farms, and whole villages testify that most beach areas will not support permanent settlement.

Human beings must always be ready to move on the beach. Two kinds of retreat are necessary. One is the rapid flight before an approaching storm. The second is periodic moving to escape the retreating shoreline. Only the smallest and lightest dwellings, those that resemble tents, could be considered movable on the typical short notice given before a hurricane. Major storms deny safety to all permanent buildings other than impossibly expensive monoliths or bunkers. But these, by their very permanence, must not be close to the water, or shoreline retreat and island migration will defeat them.

Modern design and technology are prepared to offer any number of quickly movable structures for beach residences. Just as "nomadic" furniture has been popularized at all levels from the architect to the handyman, so small nomadic housing might come of age, especially on the beach of the future. A throwaway house of treated cardboard would recognize the inevitability of hurricane loss, minimize the battering-ram debris in a stormy sea, and create less political demand for protective work. In this era of scarcity, removable houses are even better. The building industry offers a supermarket of modular housing, often boasting of how it can transport whole rooms ready-made to a building site. This housing is still too inflexible for a beach, but if the same industry would concentrate on lightweight, biodegradable, nomadic housing, it would be doing a great service for beach lovers. Lightweight modular components are presently available or possible in foam, fiberglass, cardboard, and inflatable plastic—all of which lend themselves to being moved in pieces or as a whole. Seasonal removal would allow insurance companies to write cheap policies for the calm weather of April through July. Movable homes would follow the lead of boat owners, who for centuries have hauled their craft out of the water in the fall and returned them to their moorings in the spring.

People who presently own beach homes and businesses, or those who desire them, will not settle for the small nomadic dwellings we have just described. Those who insist on greater space and comfort must begin by accepting the risk of destruction by storm. After that, they still protect themselves from shoreline retreat by limited mobility.

Houses and stores can be built so that from time to time they might be stepped backward just as the beach itself steps backward.

We're not leading to an endorsement of mobile homes on beaches. Most mobile homes do not fit the movable category, since they are scarcely more movable than a frame house or most modulars, and they are moved only by commercial haulers. What's more, in a storm mobile homes are prone to capsize and become lethal debris. Given the frequency of coastal storms, relying on mobile homes as a response to shoreline retreat exposes all other buildings to certain and unnecessary danger. The new generation of "motor homes" offers a much safer beach residence than a mobile home, though neither, parked in great metallic herds, improves the beauty of the beach.

Moving a building becomes nothing more than talk if lot lines are not also movable.

Utilities. Community electric, water, and sewer systems can be built in such a manner that they can be disconnected from dwellings just as they are disconnected in motor-home parks. In the same manner, water and sewer lines should never connect islands or even parts of a single island that are likely to be separated by inlet formation. Pipelines running the length of an island encourage people to try to prohibit inlet openings and to fortify weak areas. The utility buildings, tanks, and treatment basins should be underground or built with a very low profile and sloping sides, so that they can be overwashed and partially buried without damage.

A fine example of this kind of development is the Bay Marchand, Louisiana, oilfield. Here Shell, Phillips, Gulf, and Chevron oil companies have established wellheads and small holding tanks in an area where the shoreline is retreating some sixty feet a year. Bay Marchand is on an arm of the Mississippi Delta that is no longer active, and the fine sediment of the shore is settling. As it settles the sea advances. In a compromise with the sea, the oil companies designed their fields to become offshore fields as the water inundated them. Wellheads were made extremely durable so they could pass right through the surf zone. The oil companies chose this design strategy when they realized that the cost of an armored shoreline would be prohibitive.

Roads. Every time a beach road is washed out it is rebuilt higher and more solid than before. Exactly the wrong approach. If a road is tough enough to stop even short-lived erosion, it has become a seawall focusing wave energy in a small area and causing fundamental erosion. Even if erosion is not severe, a road that halts the cross-island flow of overwash sand severely interrupts natural beach maintenance. Roads parallel to the shoreline should be as short as possible. Rather than running the length of the beach, as they do on many barrier islands, they should run perpendicular to the shoreline if at all possible. But they should never run straight and flat into the dune fields, inviting flood tides to use them as overwash passes. If beaches must be accessible by car, it is preferable (though considerably more expensive) to have several bridges or causeways from behind the beach rather than one or two at each end connected by a road paralleling the waterline like an uninterrupted bulkhead or seawall.

Dr. David Duane, once with the U. S. Army Corps of Engineers, has suggested that gravel parking lots on Cape Cod would create less expense and debris when the endlessly eroding bluffs caught up with them. As the lots fell to the beach, the gravel, of course, would blend with the glacial till of which the Cape is made. Elsewhere the National Park Service is moving toward cheap, short-lived roads that can yield easily as the beach responds to storm, wind, and erosion.

Bridges. Inlets on barrier islands open and close quickly, and migrate rapidly up and down the coast. Filling a new inlet to re-establish the status quo is often difficult and expensive, sometimes costing more than a new bridge. The task of inlet filling also requires a large supply of precious sand. Bridges built over an existing inlet, however, may be quickly undermined. In other cases the inevitable inlet migration leaves the bridge spanning nothing but dry sand. If at all possible, ferries should carry people across inlets. If not possible, bridges must be movable. The military, to accommodate its own mobility, has developed a number of bridges that can be taken down and put up in less than a day's time. One telescoping bridge rides on a self-propelled crawler. Compared to the constant maintenance costs of a permanent bridge, these movables are a great bargain, especially since they require no jetties to fix the position of the inlet. Movable bridges might not handle the great volume of private cars that visit the beaches. Private beach areas as well as public are best served by mass transit, a system that could generate income for the local economy as well as relieve the monumental traffic jams of a sweltering summer day. For years the public has accepted mass transit to the beaches served by ferries. Mass transit on land should be just as feasible.

Though each beach will inevitably make its own demands on land use and development, the possibilities and general policies we have described can allow us to visualize a typical high-energy beach—say, a long, narrow barrier island. From the bay side several permanent bridges approach the island, stepping carefully through the marshes on a minimum of supports. They end on relatively high, wooded areas far back of the primary and secondary dunes. Not far from the bridgeheads are a few low-profile motels, and beyond these a few condominiums and apartments. Between these developed bridgehead areas there are only gravel and packed-sand roads that wind in and out of small dunes on the back beach. During the summer along the secondary dunes are many scattered, small, movable cottages. Low pillbox-like structures here and there are access rooms to sewer, water, and electricity lines. In several places the primary dunes are bridged by boardwalks and tracks for miniature cog railways carrying cars like those used in airport terminals and zoos. The public transport begins near the bridgehead and ends on the public beach.

Some of this is futuristic thinking, but we could begin now to adapt presently developed areas gradually to movable development. In high-risk zones or where nonconforming use permits exist, developers and homeowners might be notified that in the event of destruction, a new building and zoning concept would apply. While it would be impractical to do away with private-property rights, fixed lot lines might be discouraged or banned—in recognition of the likelihood that nature will alter them even if man does not. As in many condominiums, property owners would own the land in common and the space within their houses privately. A person would own only what he or she is capable of controlling. The overall title to a development area would have to be vested in a local government, a cooperative, a landowner's association, or a development corporation. Within each development area there would be a certain number of movable living modules allowable—determined by health and safety requirements. Developers would also have to provide a plan for accommodating sand-dune movement and emergency evacuation in the event of storms. This differs little from a zoning code which allows only a certain number of houses per acre.

If the sum of all the acreage became too small to support the number of dwellings present, someone would have to move. Making this choice, even if normal attrition did not take care of it, would not be as difficult as watching whole developments disappear in a storm or into the channel of a migrating inlet. And movable houses, after all, could be sold to someone on another beach area or island. The payoff of movable shoreline development is that when and where the natural forces become too great for man to adapt himself rather than modify nature, the shore can be deserted with no ill effects for either party.

Our plan for environmental harmony might best be characterized as a plea for making what coastal engineer David Bella calls "good mistakes."

A good mistake is the kind of mistake we wish the past had made—mistakes that were not irreversible, that did not limit our choices today. Building Atlantic City or Miami Beach is an almost irreversible mistake. Our present choices on those beaches are limited and unhappy. This gentle, future-minded approach to planning calls for five commitments, according to David Bella:

1. Protect the choices of future generations by avoiding large-scale, irreversible changes.
2. Protect the flexibility and diversity of nature so it can adjust to our own mistakes and to future mistakes.
3. Preserve a variety of environments from which future generations can learn and choose.
4. Use technology carefully and do not make decisions which require complex control and perfection. Preserve for the future the choice of discarding today's technol-

ogy rather than having to perpetuate it in order to avoid crisis.

5. Reward people who reveal mistakes and admit their own failures and learn from them. Do not encourage the camouflage of mistakes through corrective technology, especially if the problem could more safely be solved by admitting and removing the mistake.

Bella's approach is particularly appropriate to beaches. Civilization on the beach, no matter how grand or sophisticated, exists on the frontier, on the edge of wilderness.

Only a purist maintains that true wilderness can show no mark of human life. Even the map and survey lines which outline the wilderness boundaries of Alaska's Mount McKinley or the Grand Tetons or the Great Smokies taint those areas with mankind's work. Wilderness is not judged well by the eye alone. The key to wilderness on the beaches is what form of energy controls life there—energy directed by the self-willed designs of man, or the energy expressed through the infinite varieties of natural forces. Most beaches, because of the tremendous energy exerted through nature, continue to re-establish their wildness despite man's settlements. And the narrow strand that struggles to preserve itself is only the edge of the most vast wilderness of all—the great moving oceans that cover two thirds of the earth's surface. Whatever people do at the beach—fish, swim, or sunbathe—intuitively most come to do these things in the presence of something elemental and wild. Those who are healthy enough enjoy testing themselves physically—on surfboards, in boats, or in bathing suits—against the water or the wildlife in it. The wilder the better.

But when the wildness of the world rises up and strikes us, we are outraged. We forget our deep and mysterious kinship with it. We forget that to love something does not mean to tame it. We react like Captain Ahab after he has lost his leg to the giant white whale. "I see in him outrageous strength and inscrutable malice sinewing it," Ahab says. "That inscrutable thing is chiefly what I hate." And like Ahab, we set out to punish and conquer the forces that seem to have betrayed us.

Like the whale, the shoreline and the ocean can be conquered by humanity. But unlike the conquest of a single species, this conquest can only be temporary. The tides and waves breaking on the world's shoreline cannot be tamed, even with all the energy and materials that the continents can supply. To the extent that we do subdue the shoreline in our limited fashion, we subdue ourselves—at least some part of ourselves that, like the ocean and the beaches themselves, is still mysterious and unpredictable and moving.

Epilogue: It's Not All Over Now

A YEAR AFTER the first edition of this book conservationists, congressmen, governors, and President Jimmy Carter declared 1980 "The Year of the Coast." We certainly don't take credit for, although we hope we were part of building, the understanding of the beaches that blossomed in that unusual consensus that something had to be done right away.

The legal centerpiece of the Year of the Coast became Congressman Phillip Burton's Barrier Islands Bill (later rewritten as the Chafee-Evans Bill). California's Burton, together with thirty cosponsors, had introduced a bill that would catalog the nation's remaining portions of undeveloped barrier islands and prohibit all federal aid that could stimulate development. No water, sewers, roads, insurance, or erosion control. The secretary of the interior would designate protected areas and administer the law.

As the new year began, Ronald Reagan came to Washington and introduced the man who would be his secretary of the interior and whose support for the bill would be crucial. James Watt brought to the capital his fame as a lawyer for the Sagebrush Rebellion that had opposed environmental regulations, expanded parks, and wilderness out West. Early in his tenure Watt said future generations were of little concern since God would end the world soon anyway. As for environmentalists, he said, "For the most part all they are interested in is membership dollars and headlines."

Realtors and the National Association of Homebuilders sensed support and lobbied heavily for continued building subsidies. Secretary Watt and President Reagan hesitated. Not only were they against subsidies, at least for social programs, but they were trying desperately to wrestle with a $100 billion budget deficit, the largest in history. The growing deficit embarrassed the president's conservative allies, and some of them felt the Barrier Islands Bill could eliminate one of the "giveaways" that was making taxes so high and swelling the deficit.

The bellwether of these allies was Howard Jarvis, who had won fame as the promoter of the antitax Proposition 13 in California. Jarvis got his National Taxpayers Union to run a full-page ad in the *Washington Post* listing a priority agenda for the Reagan administration and its new backers in Congress. Prominent on that agenda was "Safeguard natural protective barriers against storms, beginning with America's barrier islands (H.R. 5981)." Just when some of the president's backers saw in James Watt a man who would silence the voice of conservationists in the administration, here were others forging a new bond between the conservation movement and conservative politics.

Opponents of the bill thought if they could not defeat its philosophy, they might at least delay and drown it with artful confusion. They recruited a small group of scientists to rewrite the definition of *barrier island.* These scientists said only islands less than 7,000 years old were true barrier islands. Since most barrier islands include portions of older islands, this definition would have fatally stalled the bill in endless scientific debate. The islands, regardless of age, would continue to move while the opposing scientists sifted sands and analyzed core samples in the laboratory.

The bill's backers were prepared for the challenge. In March of 1981 coastal geologists meeting at the Skidaway Institute in Savannah, Georgia, had produced a widely read position paper on the importance of controlling beach development. Most of their conclusions match those we have put forward in this book. By July of 1981 the momentum was with conservationists. Congress slipped a provision into the president's budget that cut off flood insurance for undeveloped barrier islands by 1983. By this point statistics were showing taxpayers providing a $400 subsidy to every coastal insurance policy. The Skidaway statement warned of worse economic disasters. Despite the failure of the attempt to redefine barrier islands even Secretary Watt climbed on the bandwagon. He called the bill "an opportunity for substantial savings in federal expenditures." In October of 1982 President Reagan signed the bill, which had passed the House of Representatives by a vote of 399 to 4.

In the meantime the Reagan administration had almost wrecked the Coastal Zone Management Act. The first budget recommended no funding for the federal or state CZM programs. At the same time Secretary Watt proposed to open almost the entire continental shelf to oil drilling regardless of its onshore consequences. Without operational funds, states would be hard-pressed to analyze federal leases and provide a sound scientific underpinning to their right of review and challenge as guaranteed in the federal Coastal Zone Management Act. The Reagan people were counting on $18 billion from offshore leasing to help them tame the budget deficits.

Congress approved funding anyway, though reducing it from over $100 million a year to $40 million. To provide future funding it proposed an Ocean and Coastal Resources Management Fund to be financed with less than 5 percent of the coveted offshore revenues. Secretary Watt could not object too strongly since his home state of Wyoming was receiving millions of dollars a year from its 50 percent share of federal mineral-leasing revenues.

Coastal Zone Management programs will go on, and some 623 miles of barrier islands from Texas to Maine will be protected. This is only a beginning, not an end, of the work to control the exploitation of the beaches. If the Barrier Islands Bill saves the federal government an estimated $6–11 billion over twenty years on such a small amount of acreage, we can hardly imagine how much is at stake in areas where government and private investment is already intense.

Our attention must now focus on the thousands of miles of mainland beaches, already developed islands, and island areas exempted from protection for political reasons. Existing development at Miami Beach, Monmouth Beach, Ocean City, South Padre Island, Malibu, and many other areas will cost many lives and billions of dollars in the near future. The pressure to "stabilize" these beaches will threaten both developed and undeveloped beaches nearby. America has just begun to think about and act on this problem.

Since the appearance of the first edition of this book in 1979, we have been gratified by the support for its approach to coastal problems. We said in our acknowledgments that the book's value did not issue from the virtues of its authors, even if we do not count ourselves among the scoundrels. We said "we are the conduits through which the people we have known may talk to others who will read this book."

Two years after we wrote that opening, coastal geologists issued their extraordinary proclamation from the Skidaway Institute. "Saving the American Beach" begins by saying, "New approaches to the management of the American shoreline are urgently needed to preserve our recreational beaches for future generations." This statement declares, as we have, that the natural beaches are not threatened by shoreline retreat but that engineered defenses "seriously degrade the environment and eventually help to destroy the areas they were designed to protect. . . . Finally we become locked into a dilemma of costly counterproductive measures wherein the more we do the worse the problem becomes."

The scientists, representing two-thirds of the coastal geology community, list numerous examples and recommend specific solutions. This book has been an outlet for much of that material. The scientists conclude with a demand for an honest and professional rethinking of our approach to coastal management.

We sense in the Skidaway statement a historic new beginning. Some states have already begun over again in their coastal planning. Early in 1983 Maine became the first state to declare that it would undertake no more shoreline defense projects. Others are moving in that direction. These states will provide a good yardstick against which to measure a state like Louisiana, which intends to continue defending the present shoreline.

Despite the holdouts, a new vision of the coastline has begun to develop in America. Coastal geologists issued the first Skidaway statement, confirming our conclusions about the future of our beaches. The authors of this book, with Dr. James Howard of Skidaway, convened a second conference to consider how Americans should respond to the crisis. This time the experts came from several key disciplines. These experts accepted the geologists' analysis of the problem and concluded that with very few exceptions the only good response would be to retreat from the shoreline.

This conclusion might seem idealistic, but most of the

experts were people who deal in the realities of day-to-day economics, law, development, politics, and engineering. The introduction to this second statement summarizes their findings:

> Sea level is rising and the American shoreline is retreating. We face economic and environmental realities that leave us two choices: (1) plan for a strategic retreat now, or (2) undertake a vastly expensive program of armoring the coastline, and as required, retreating through a series of unpredictable disasters.

The experts knew they had to back up this strong conclusion with practical recommendations. After many hours of hard work, they forged a consensus rare for such diverse points of view. They offered a long list of solutions, some practical for only a few sites, others for many. The point, they noted, was that we have many ways to solve our problems.

Excerpts from the second Skidaway document follow:

> Since almost all communities must ultimately rely on achieving protection by integrating development and the natural system, stabilization must be severely restricted. This fact and the changing scenario created by rising sea level means we must also continually reevaluate the costs and benefits of existing structures. . . . We must be ready to remove, revamp, demolish, or relocate some major structures when their existence becomes more burden than benefit to either the natural environment or the public *per se*.
>
> HOW TO RETREAT
>
> Since the general public interest is at stake, government must take the lead. In "coastal high hazard

areas" we recommend that the following measures be considered:

FEDERAL GOVERNMENT

1. End all federal expenditures, direct or indirect, in support of private coastal development. Require private coastal development to pay its full cost.

2. Replace economic incentives for private development in high risk areas with incentives to relocate and build in other areas.

6. Encourage research in new technologies for managing beach areas, especially inlets and navigation channels, without disturbing natural processes.

7. Provide special tax incentives and disincentives to limit development in the units of the Coastal Barriers Resources System and V Zones (undeveloped lands and high hazard zones respectively). . . .

b. Allow tax deductible gifts (to govt. or conservation groups) with the right of the owner to use improvements until damaged by erosion or storms.

e. Treat gains on property in high risk areas as ordinary income, rather than as capital gains.

g. Repeal the deduction for interest paid on loans for properties in high risk zones.

i. Give preferential tax treatment to profits made on sales to public bodies or conservation groups.

STATE GOVERNMENT

(many recommendations parallel federal actions)

6. Create a property transfer tax to fund acquisition of important coastal resources, public beaches, and beach access, as already done in Florida and Massachusetts.

10. Require developers and real estate agencies

marketing property to disclose in writing the risks of being in the high hazard areas, including the costs associated with such risks during the expected life of their building.

11. Require, when recording each change of ownership or new financing, [that] a current plan be filed showing the lot lines, location of buildings, and the shoreline location. Deed descriptions might note specific risks of hazard zones.

12. Require a successful applicant for a permit to rebuild in a high hazard area to waive rights to petition government for public aid when future damage occurs.

LOCAL GOVERNMENT

2. Assign a non-conforming status to high risk uses of land just as zoning codes consider certain uses non-conforming. Regulations could prohibit non-conforming uses from being rebuilt after a certain level of damage has been sustained.

3. Require new subdivisions to set aside lands in safe areas for those who must retreat from the shore. . . .

4. Remove or require the demolition of structures that become a threat to public safety, including seawalls and other structures in the surf zone, and high risk buildings.

8. Develop zoning provisions that have special standards for areas of unstable beaches, including a "floating zone" in which zoning designation and standards move with natural features such as mean high water, dune, or vegetation line.

9. Levy special impact assessments on risky devel-

opment to provide a reserve fund for buying out damaged properties.

PRIVATE SECTOR

1. Develop innovative technologies to adapt to changing public policy with emphasis on new modes of sand by-pass, inlet maintenance, and residential construction.

2. Real estate organizations such as the National Association of Realtors and the National Association of Homebuilders should educate their members about the need for new policies and about development patterns that can minimize the effects of new regulations.

3. Professional appraisers and economists should develop standards for assessing the effects of new policies on property values.

CONCLUSION

Our creativity can serve us as constructively in these new directions as it tried to do in building defensive structures. We already have the technical, legal, scientific, and economic tools to help property owners, to protect local treasuries, and to assure the public that its valuable beaches will be preserved and available.

Most of these tools have been available for many years. Our faith in builders and inventors and our preference for winning a battle once begun have made us ignore these less spectacular and tangible tactics. Repeated and costly defeats on the ocean beaches, however, should convince all but the most reckless that the time is overdue to build a new strategy based on our new understanding. Only a foolhardy strategist eliminates retreat as an option, and it is even more foolhardy not to learn how to turn retreat into victory.

Copies of the entire statement from the second Skidaway Conference can be obtained from: Director, Skidaway Institute of Oceanography, P.O. Box 13687, Savannah, Georgia 31416.

CONVENORS: James D. Howard, Marine Geologist, Skidaway Institute of Oceanography; Wallace Kaufman, Real Estate Appraiser and President of Saralyn, Inc., Pittsboro, NC; Orrin H. Pilkey, Coastal Geologist, Duke University.

PARTICIPANTS: Sarah Chasis, Attorney, Natural Resources Defense Council, New York, NY; Robert G. Dean, Coastal Engineer, Department of Coastal and Oceanographic Engineering, University of Florida, Gainesville; Col. Paul S. Denison (USA, Ret.), Consulting Engineer, Henry von Oesen & Associates, Wilmington, NC; David R. Godschalk, Professor of City and Regional Planning, University of North Carolina at Chapel Hill; Peter H. F. Graber, Attorney, Coastal Law, Greenbrae, CA; H. Crane Miller, Attorney at Law, Washington, D.C.; Robert A. Morton, Coastal Geologist, Bureau of Economic Geology, University of Texas at Austin; Sharon Newsome, Director of Legislative Affairs, National Wildlife Federation, Washington, D.C.; James Nicholas, Economist, Joint Center for Environmental and Urban Problems, Florida Atlantic University, Boca Raton; Walter D. Pilkey, School of Engineering, University of Virginia.

Checklist for Buying or Building on the Beach

WHEN YOU GO to the beach to buy property, you are on unstable ground. Not only is the sales pitch more aggressive, but the salesperson who possesses a native's knowledge, or for that matter any knowledge, of the beach is a rarity. Even rarer are the salespeople who will tell you what they do know of erosion, hurricanes, inlets, or other natural hazards. The simple checklist we have made will help you navigate at least some of the market's shoals.

On the East and Gulf coasts your first decision is to choose between mainland and island locations. On the West Coast you will be choosing between bluffs and cliffs, coves and sand spits. Once you locate your beach, you will probably be faced with many homesites—high and low, wooded and bare, mid-island and waterfront. Finally, having chosen the lot, no matter how safe, you must still build for the special environment that is the beach.

When you know the answers to the following questions, you can make a wise and environmentally sound investment.

Decision 1: What Type of Shoreline to Live On

BARRIER ISLANDS

1. How wide is the island? A wide island is not as easily breached by inlets or flooded by storm surge.
2. What is the island's elevation? The higher the island; the greater the storm necessary to flood it, and the more secure it will be against rising sea level.
3. Is there natural shelter from high winds on the island? A forest canopy sheds some wind and blunts its force.
4. Does the island have a fixed bridge or a drawbridge for access to the mainland? A fixed bridge cannot jam open.
5. How many bridges provide island access? More bridges mean more escape routes. Bridge aproaches should be high enough to stay above flood waters. Escape routes from towns such as Nags Head, North Carolina, and Port Aransas, Texas, are nearly useless because they are blocked early in the progress of a storm.
6. What has been the storm history of the island? How have houses fared in past storms? Remember, local residents, especially real estate agents, tend to have short memories.
7. Does the island have a history of inlet formation during storms? Will inlets block roads and cut off services?
8. Is the fresh-water supply likely to be cut off during a storm? On some islands, waterpipes cross bridges or the sites of possible inlets and could be ruptured in a storm. Loss of water supply can be long-term or even permanent.
9. Where does the sand for the beach come from? If the beach is artificially nourished, chances are the sand

will wash away and future nourishment costs are likely to fall heavily on local taxpayers. If beach material is derived from nearby bluffs or cliffs, attempts by cliff-dwellers to protect themselves might starve the beaches.

CLIFFS AND BLUFFS

1. How "solid" is the cliff? Well-consolidated cliffs (solid rock) are far safer than gravel or shale cliffs. A steep bluff face in unconsolidated material is *certain* evidence of rapid erosion.
2. Is the beach littered with evidence of previous cliff failure near your site? Piles of rubble at the base of the cliff are sure signs of such failure. A clean, smooth beach is little insurance, however, since the waves that eroded the cliffs might also have carried away the debris.
3. If you can see layering in the rocks of the cliff, in what direction does it incline? If the layers dip toward the sea, this provides a natural "slide" for cliff failure.
4. What type of rocks make up the cliff? A layer of shale or an unconsolidated layer between hard-rock layers both spell potential disaster for the homeowner living above.
5. What has been the history of bluff failure in the area according to local residents and public records? Check the recorded maps for land near the cliffs and see if the older ones show streets or lots that have disappeared.

ALL BEACHES

1. Is the shoreline growing seaward or eroding? This question should be answered for a period of at least

twenty years. What happened last year or three years ago may be temporary. Signs of rapid erosion are: 1) tree stumps and clumps of mud or black peat on the beach; 2) steep scarps (bluffs) in the first dune, especially with protruding plant roots; 3) a short distance between the high-tide shoreline and the first dune; 4) evidence of previous attempts to stabilize the shoreline—remnants of groins, seawalls, and bulkheads.

2. How did the beach respond during the last big storm? Was it cut back significantly? Did most of the beach return within the next few months? Were houses lost or damaged? Were cobbles from the beach thrown into houses?
3. Is erosion likely to increase in the future due to man's activities? If your neighbors are bulkheading their property, yours may begin to erode. Large seawalls and groins and jetties usually rob sand from downdrift beaches.

Decision 2: Choosing a Homesite

1. What is the elevation of your lot? The level of the 100-year flood is a minimum. Local planning officials can show you a map of the predicted 100-year storm-surge spread. You can build in the flood area, but you will pay more for insurance, you'll have to build a more expensive house, and you are in grave danger. If no one can tell you where the flood area is, that is good reason to look for another lot in a community that cares.
2. How does the terrain protect you from wind and waves? A forested area is generally safest, not only

because it breaks wind and wave energy, but because forests do not exist on sites of frequent salt-water flooding or overwash. High dunes between the lot and the sea may help. Hilltops are better than low areas.

3. If you are on a cliff or bluff, does your deed or do other records show anyone owning land between you and the water? This could be the evidence of a lot like yours that has already eroded.
4. Does a storm subject the area to overwash? Common sense plus a look at the soil should answer the question. The presence of seashells in or near your lot should make you suspicious.
5. Has the area been stable for a long time? If real soil has developed, you can usually be assured that the area is relatively unaffected by storms. If beach-like sand is present, beware.
6. Has the lot been artificially elevated? Try to avoid such sites. The presence of shells stained black or bleached white may tell you the sand has been pumped up out of the lagoon or bay. Compare these shells to beach shells.
7. Are local dunes stabilized? Moving sand dunes can bury roads and whole houses. Naked sand dunes are usually moving dunes.
8. Do your neighbors have bulkheads, groins, or seawalls, or are they building them? If so, try to avoid that area. One person's grab for sand will cause others to follow. Seldom can you be the odd man out without losing much of your lot. If you follow the crowd, you're likely to spend more money on defenses than on your house.
9. How far is the lot from a present inlet? A good rule of thumb is to keep at least a half mile away. Some inlets migrate over a mile.

10. What happens to the lot in the off season? Some lots between dune rows may become marshes during wet seasons. Some developers have created lots during the summer months when the beach extends farther out to sea.
11. If the lot is near the lagoon or bay, has the shoreline been altered during construction and development? Salt-marsh removal, destruction of forest, and removal of stumps from beaches all increase shoreline erosion rates during storms. Remember, big waves form and erosion rates can be high in "quiet water."
12. If the lot is on an eroding beach, is it sufficiently deep to allow you to move the house back when threatened? Check with local zoning ordinances for minimum lot size.
13. Are adjacent houses well constructed or likely to be? Weakly built houses may disintegrate or float off the foundations and batter your house to pieces. A good rule of thumb is never to build near a mobile home or a mobile-home park. Mobile home "tie downs" are of limited value, and most of the porches and cabanas are not tied down.
14. Are neighbors the kind of people who will carefully prepare for a storm? Boats, chairs, flagpoles, firewood—anything that floats—can damage nearby property.
15. Do local regulations and politics assure long-term protection for dunes and forests? Look at the professions of local officials and their record in granting exceptions to the law.
16. How fresh is the fresh water? Are there strong, well-enforced regulations to protect ground water from pollution by septic tanks and cesspools?

17. Do local regulations keep beach buggies off the dunes? Vehicles and even heavy foot traffic can destroy dune vegetation and set the sand in motion again.
18. Is the property suitable for septic tanks? Frequently a layer of marsh mud, buried by natural proccesses or by man, prevents septic-tank effluent from percolating down. If a local health official approves an unsuitable site or a marginal one as a favor to you, he has probably done it for many people, and the neighborhood could soon be contaminated and billed for expensive central sewage collection.
19. What building codes are in effect and how old are they? If building codes are new, neighboring older houses may be poorly constructed and fall apart in a storm.
20. Are building codes enforced? Frequently code enforcement is light. Again, an inspector who does you a "favor" has probably done enough favors to assure a dangerous neighborhood.
21. Are you prepared to take all the risks of partial or total destruction for the sake of the benefits of your location?

Decision 3: Choosing a House

1. What weather conditions do you expect the house to face? Design for the worst, and work in your other needs around safety.
2. Is the house well built? Structural members should be attached to each other strongly. This will generally add 10 per cent to building costs. A beach house should be built so that it can be picked up, turned upside down, and shaken without falling apart.

3. Is it feasible to strengthen an older house? How much will it cost to add bracing and metal connectors and foundation tie-downs?
4. If the house is on pilings or posts, are the floors securely fastened to each pile? In some hurricanes, houses on pilings have disappeared while the pilings appeared to be untouched.
5. What is the roof style? A hip roof sloping in four directions is more secure than a gable roof sloping in two. The more planes the roof is broken into, the less lift the wind can generally apply. Reducing gables also reduces vertical wall area for the wind to push on.
6. Is your house designed not only to withstand wind but the impact of floating debris? A minimum standard should be resistance to the impact of a 1,000-pound object moving at 10 miles per hour.
7. How many large glass areas do you really need? Big windows frequently explode in sudden pressure changes during a storm. Can you replace large picture windows with numerous smaller windows arranged to give you a greater variety of views with as much light as a single larger window?

Where to See the Beaches Moving: The Most Beautiful and the Most Beleaguered Beaches on Three Coasts

WHERE WOULD WE take you if we had unlimited access to boats, planes, and four-wheel-drive vehicles, and time to walk wherever we wanted? We would take you to those places where you could see the beach in motion—or at least where the evidence speaks clearly. Some of our stops would be amid the unspoiled scenery of Sierra Club calendars and coffee-table books. But we would also stop at some familiar recreation areas and the shores of big cities like Miami Beach.

Our tour is unabashedly biased—a tour of wilderness and of human folly, but including a few stops where development exhibits wisdom. We leave out some famous beaches that are perfectly nice places to visit. We are not very keen about lying in the sand, riding surfboards, or bobbing gently on rubber rafts. Readers can find beaches for these activities with little help. To us the beach is a place of revelations, a book of changes. We hope that our tour will change the way you look at yourself and at the edge of the land on which we live together.

The Rachel Carson Seacoast. This is not one place but twenty-three islands along the coast of Maine acquired by the Nature Conservancy for their scenic value and rare plants and animals. (The Conservancy is a national nonprofit conservation organization which acquires important natural areas and manages them in the public interest.) The islands of the Rachel Carson Seacoast are scattered along the entire coast of Maine, and all provide good examples of the geologically new shoreline, a rocky shoreline recently drowned by rising sea level. Our favorite, Maine Island, is not secure in the Conservancy's seacoast holdings, but its southern portion should be soon. Great Wass Island, off the eastern Maine town of Jonesport, contains huge bogs and fine forests. The shoreline is beautifully rugged. The U. S. Fish and Wildlife Service says the black duck population here is the nation's third largest. The beach ledges are important places for seals to rest.

In between rocky fingers of island the water is unusually deep, the result of the glacier gouging valleys into fiords. This is a very slowly changing coast with no sandy beaches, and a tide range of eleven to thirteen feet which means the rocky shoreline has a broad expanse of tidal exposure where marine life is plentiful. For information on Great Wass and the Rachel Carson Seacoast, write: Nature Conservancy, 51 Chapel Street, Augusta, Maine 04330.

Acadia National Park. Here is one of Maine's few sandy beaches, a pocket beach typical of rocky glacial coasts. The sand is glacial debris washed into a sheltered cove. Mixed with it are an unusual number of shells. These are mainly barnacles and sea urchins beaten off the adjacent rocky headlands. The wild beach area has good camping facilities and is heavily visited. Don't expect to swim long, if at all, in the cold water. Access from Route 3 south of Ellsworth. Consult Park Guide for location of Sandy Beach.

Higgins Beach, Maine. Beach development in Maine is generally light because severe winters make large motels and condominiums unprofitable. Nevertheless, the lure of sandy beaches is so great that even the most fragile are sometimes covered with houses. Higgins Beach is one of Maine's rare sandy barrier spits, more properly called a bay mouth barrier. The half-mile-long spit is some 500 to 1,000 feet wide, built of sand pushed up off the seafloor. Sea level has caught up with the spit and is pushing it backward at the rate of one to two feet a year. The developed half of the spit is covered with cottages and each has its own seawall, mainly walls built with Maine's bountiful boulders. Although the present houses are in danger, and despite a 1978 storm which damaged a small inn beyond repair, developers have plans for houses on the remainder of the spit. Looking from the spit to the nearby rocky headlands makes it clear where the safest place to build might be.

Access by Route 207 south of Portland. Limited parking.

Old Orchard Beach, Maine. Every state has its inevitable beach honky-tonk. This one reveals the irony of destabilizing one of the nation's rare, stable beaches. For reasons unknown to geologists, Orchard Beach has not been retreating or eroding. Nevertheless, occasional storms washing into developments too close to the water have motivated property owners to build seawalls and bulkheads. As these reflect wave energy back onto the beach, erosion will become a way of life. Had the residents thought to build their houses on poles, as many beach residents do elsewhere, the occasional storm overwash would be little problem.

Access by Route 9 from Biddeford, Maine.

Cape Cod National Seashore. The glacier which carved the hills of Maine left much of its bulldozed burden off the coast of Massachusetts in the form of Cape Cod. The National Seashore that President John Kennedy loved and promoted offers a good variety of beaches. At the southern end just east of Orleans, Nauset Beach is a long sand spit protecting the elbow of the Cape. At Coast Guard Beach near Eastham you can see a parking lot falling into the ocean. North of Nauset light the coast is dominated by high bluffs. These sand and gravel bluffs are also eroding rapidly. At the site of Marconi's radio station the Park Service has posted signs describing the erosion that has taken place since Marconi's wireless station sent the first transatlantic message in 1902. At the extreme north end of Cape Cod, near Provincetown, the bluffs yield to large, oceanfront dunes and broad sandy beaches. Large naked dunes set in motion by unwise timbering and overgrazing move restlessly across the land.

The visitor should remember that the erosion on the Cape is natural. The Cape is a temporary land mass, geologically speaking, a pile of debris dumped in the ocean.

Green Hill Beach, Rhode Island. This one-mile-long barrier beach near South Kingston is a good example of history repeating itself. The 1938 hurricane swept the island clean of houses, killed 48 people, and caused $2.5 million in damage. The destroyed houses were swept inland where they battered other houses. Little building took place during World War II and the following decade, but a 1954 hurricane still caused $1 million damage on the barrier. The town tried to zone the beach to prevent building, but in 1965 an "oversight" invalidated zoning and the barrier is now developed with modern houses. Storm tides have often risen higher than the beach's low dune ridge. Residents have tried a variety of methods to build up dunes and stop erosion, including barriers of trash and old Christmas trees. Most of the houses on Green Hill Beach will be destroyed in a major storm as the barrier loses its dunes and the beach flattens out under wave attack.

Fire Island National Seashore, New York. The 118-mile-long backbone of Long Island is a ridge of glacial debris. The hilly north shore of the island, facing Connecticut across Long Island Sound, is much like glacial bluffs in New England. The south shore of Long Island, facing the Atlantic Ocean, is a flat outwash plain where meltwater and rainwater ran down the terminal moraine to the sea. At the edge of this outwash plain is the Great South Bay, and beyond that, separating the bay from the ocean, is part of the East Coast's first great barrier island chain. The island has been battling for its stability under the influence of jetties and groins robbing its sand from the east, where the affluent residents of the Hamptons are trying to keep their beaches and inlets in place.

Fire Island is not a particularly high barrier, and many houses have been destroyed by winter storms and hurricanes. The Park Service is now committed to letting nature have its way on the oceanfront. This is a good island to visit and revisit to watch nature and public policy at work.

Coney Island, New York. Although disguised by heavy development, filling of its lagoon, and an amusement park,

Coney Island was once the westernmost of Long Island's barrier beaches. What millions of visitors see now is a beach made entirely of pumped-up sand. This is backed by a massive seawall, on top of which is the boardwalk. It is interesting to stand on the boardwalk and imagine what this New York City beach would have been like if the island had been left to its own defenses. Today there is no practical alternative to continued replenishment of the sand. The cost, in this case, is probably justified by the millions of people who can visit the beach via mass transit.

Monmouth Beach, New Jersey. Here is a good example of a once pleasant resort community with a totally armored beach. Today a large boulder-and-concrete seawall shuts off the view of the ocean, even from some two-story houses. The wall, which looks almost indestructible, was breached in several places by storms in January 1978. On the ocean side of the wall the shallow waters reveal three generations of "erosion control" structures, mainly splinters of wood and steel groins and bulkheads. What few patches of beach remain are pocket beaches between engineering structures.

Access by Garden State Parkway and Route 36.

Ocean City, Maryland. This is an almost instant Miami Beach of the mid-Atlantic states. The barrier island is some ten miles long and three blocks wide, and since the early 1970s it has been crowded with high-rise motels and condominiums. In the first eight months of 1972 the city issued permits for $62 million of new construction. The beach has already narrowed so much that the city tried to broaden it in 1978 by bulldozing sand from the tidal beach to the upper beach. Most of the dunes that once existed on the island are now gone, leaving Ocean City a sitting target for inlet formation during storms, an event that is well-documented in this area. In fact, the inlet that separates Ocean City from the mostly wild Assateague Island to the south, only opened during a 1933 hurricane. The Corps of Engineers built jetties to stabilize that inlet, and in the process they sealed the fate of Assateague's northern end, which has eroded so rapidly that it detached itself from the jetty on its side of the inlet. To see what the beach at Ocean City was once like, travelers should visit the National Seashore at Assateague.

Both Assateague and Ocean City are reached by taking Route 50 east from Route 13.

South Bethany, Delaware. Here is barrier island development kept alive by the security of federal insurance programs. The first row of beach houses disappeared in the 1962 Ash Wednesday storm. A number of houses are now leaning into the surf zone and could be destroyed at any time. A hurricane, which could be much more devastating than a winter storm, might destroy much of the oceanfront development. Nevertheless, new houses are being constructed. For a few extra dollars the owners can buy flood insurance. The main cause of erosion here is the gradual rise of sea level. As this island loses land on the front side, developers are filling the marshes on the back side.

Virginia Coast Reserve, Virginia. An unusually beautiful group of barrier islands saved from development when the Nature Conservancy stepped in, even creating a dummy corporation to buy a key island from an anti-preservationist who said he would only sell to a developer. Of the thirteen islands, one has disappeared in a storm. The others, higher and well forested in many cases, contain many opportunities for solitude and wildlife observation. Information from Virginia Coast Reserve, Brownsville, Nassawadox,

Virginia 23413. Access by local road from Route 13. Island visits by boat only. No regular ferry.

Shackleford Banks, North Carolina. On the north-south Atlantic Coast, Shackleford is one of the rare islands that runs east and west. It is part of the new Cape Lookout National Seashore. The island offers some stretches with well-developed dunes maintained by windblown sand, and other stretches that are maintained by periodic overwash. The midsection of the island has a haunting forest of bare cedar trees, buried and then re-exposed when sand dunes moved over a former maritime forest. Behind the ghost forest is a dense and beautiful woodland full of birdlife and clear pools. Access by private boat only.

Wrightsville Beach, North Carolina. This is an old resort which has grown bolder in its development. Its time has come for major disaster. The island has already lost two rows of houses in hurricanes. The beach has been replenished twice, and the sand disappeared faster the second time around. The remaining dunes frequently show a scarp where they are eroding. Local government is financially strapped to maintain the beach and stands at the verge of committing itself to a seawall. The Holiday Inn sits on the site of an old inlet and is known to beach experts as the "Holiday Innlet." Here is a southern beach on the verge of following New Jersey's example. There seems to be no other solution to saving development. Because the bay has been dredged behind the island for a harbor and marina, the island stands like a low sand pinnacle facing the ocean and unable to migrate.

Access from Route 17 and local road.

Kiawah Island, South Carolina. The Kuwaitis bought this wild barrier island a few years ago and began development amidst great protest from conservationists who wanted the island's marsh lakes and broad beaches left whole. So far development seems to be the sanest we have seen, with few exceptions. This may be because the developers were aided by Dr. Miles Hayes, a leading beach geologist, who analyzed the significance of historical changes in the island's shape. The foredunes are untouched and setback lines take into account erosion rates and overwash. If islands have to be developed, this is one of the safest. It is a broad island in a low-energy area of the coast with a good climax forest. It was once the site of an old plantation. Marshes and woods still host alligators, deer, wild pigs, and a great variety of water birds.

Access by Route 700 and local roads south of Charleston.

Hunting Island, South Carolina. No one is sure why this undeveloped state park is eroding so rapidly, but the ocean is eating up the beaches so fast that in many places the shoreline is actually the forest itself. There have been so many stumps on the beach that the Corps of Engineers, to placate bathers, mined new beach sand from mid-island and covered the stumps. Behind them they left a large stagnant lagoon in the maritime forest. The sand was too fine and disappeared. In 1976 a second replenishment from inlet land proved more successful, but the dredging took away shoals which protected the next island downdrift and erosion increased there.

Hunting Island has camping facilities and is reached by Route 21 south from Beaufort.

Sapelo Island, Georgia. So many beautiful and relatively unspoiled sea islands line the Georgia coast that making a recommendation is difficult. Most of these barriers are wide and well forested, but the width is deceptive, since each island is really two islands. The modern part of the island is a thin strip along the ocean that was once farther out to sea. These narrow modern islands have backed into the remains of islands left before the last glacial period. Sapelo Island is the site of the University of Georgia Marine Institute, where scientists have been carrying out research into beach dynamics. The island's maritime forests are full of wild cattle, deer, raccoon, snakes, and other wildlife. The fine sands of Sapelo Island spread out in broad, low-angle beaches.

Access by U. S. Route 17 from Savannah to the north, or from Brunswick to the south, then by ferry to the island. Primitive camping and guided tours.

Miami Beach, Florida. The beach has recently been replenished by a major dredging effort. New sand has covered the old groins, but the bulkheads in front of the hotels still stand as reminders of where the ocean once lapped every day and where it will lap again if replenishment ever stops. The restoration project included no dune building. The visitor should stand on the beach and imagine what the scene will be when a major hurricane comes with a storm surge ten or fifteen feet high topped by pounding waves. This island once had extensive mangrove forests and swamps on the lagoon side and an extensive dune system on the oceanfront.

Bahia Honda State Park on Bahia Honda Key, Florida. This is a typical small beach of the Florida Keys. The only sand here is made of the skeletons of marine animals. It is a light sand that forms little pocket beaches between hard limestone headlands. Unlike the headlands of New England or the West Coast, these limestone headlands are low and often unnoticeable. On Bahia Honda the beach is only a quarter-mile long. A nice place to pull off the nearby road for a picnic.

Treasure Island, Florida. Here is a heavily developed island off the Tampa–St. Petersburg area that fits almost exactly into the plot of John MacDonald's novel *Condominium,* where a hurricane sweeps across a barrier island, opening an inlet from the bay to the Gulf, collapsing high-rise condominiums. Most of the residential areas on the bay side of the island were pumped up out of the bay. The main boulevard is flooded in most heavy rainstorms. The inlets at each end of the island are stabilized and incapable of handling large volumes of storm water. The condominiums on the Gulf shore are frequently built illegally close to the water, and some slabs and walls have cracked in recent storms. The storm destruction in the novel is sensational, but certainly could happen on this very island. The disaster could in fact be worse. Several large colonies of cheap bungalows and hundreds of mobile homes would be wiped out by any overwash.

Access from the north via Indian Rocks bridge and from the south by John's Pass bridge.

Petit Bois Island, Gulf Islands National Seashore, Mississippi. This little barrier island, some eight miles off the Mississippi coast, may soon disappear. The island, like most of the islands in the Seashore, has been gradually migrating westward, accreting at that end and eroding at the other. At the western end of the island is the major shipping channel for Pascagoula, the nation's fifteenth largest port. Each year the channel is dredged the island loses seven acres of its beach. Until the loss reaches a critical state, the island will continue to be an important habitat for snow geese and many kinds of ducks.

Boat access only from Gulfport, Mississippi.

Pass Christian, Mississippi. This beach is typical of the mainland coast. The beaches here, as elsewhere, are pumped up. Even though the wave energy is low, the sand continues to disappear. The artificial beach is backed by bulkheading, which is the curb of the scenic coastal highway. On the other side of the highway is dense residential development. In Pass Christian you can still see bare foundations on the landward side of the road, remnants of 1969's Hurricane Camille testifying to the eventual fate of the new houses nearby. The older houses you see here and there survived by being built on higher ground. Along the road are scenic stops but no toilet facilities or camping to encourage visitors to use local beaches.

Grand Isle, Louisiana. This small barrier island south of New Orleans illustrates the fate of islands in the delta region once the river mouth has swung away from their location, depositing sediment elsewhere. Erosion becomes rapid. The Corps of Engineers has replenished the beach sand, but since residents would not or could not move their houses, the sand was dumped too far from its best beach position and has disappeared rapidly. This is a low-overwash island. Homes have been repeatedly devastated.

Access by Route 90 from New Orleans and then Route 1 south.

Galveston, Texas. This is the scene of the greatest hurricane disaster in American history, the 1900 storm which killed six thousand people on Galveston Island. This disaster led the town to build a large seawall, which has saved it from several storms but only at the price of losing the beach. With the exception of sand accumulating immediately behind a jetty, no beach exists in front of the seawall, although as late as 1965 you could drive a car in front of the wall. At the western end of the seawall, you can see where the natural shoreline has moved to and where it would be on most of the island were it not for the wall. Where the wall ends the beach suddenly cuts inland 150 feet.

Galveston is reached by Interstate 45 south from Houston.

Sargent, Texas. Here is a Texas beach that was built by a river delta but which is now eroding rapidly because the Colorado River, several hundred years ago, deserted a course which brought it to the Gulf updrift from Sargent. Two rows of single-family cottages have already been lost. The mud, shells, and rock that overlay this beach are washing out of the old river delta. Old delta areas like this are extremely vulnerable to both ordinary erosion and storm attacks. Reached by Route 457 from Bay City.

Padre Island National Seashore, Texas. This is the world's longest barrier island, most of it now preserved as public property. To savor the bleak solitude of these semidesert beaches, we recommend Little Shell and Big Shell Beaches a few miles south of the Malateite Ranger Station. Access is by foot or by four-wheel-drive vehicle over soft sand. A park road will get you to the Ranger Station from Corpus Christi. At Big Shell and Little Shell the foredunes are high and the beaches are broad. On the back side of the island are the vast wind flats of salty Laguna Madre. On these flats new land will one day rise as the island retreats before rising sea level. Bring plenty of water, binoculars for watching the abundant bird life, and a tarpaulin or poncho to stake out for shade. There are no trees near these beaches.

South Padre Island, Texas. As if to signal what might have happened to the entire island, this resort colony of motels and condominiums sits on the rapidly eroding beaches made of materials brought to the coast by the old delta of the Rio Grande River. The southern part of town has been accreting behind the jetty for the Brownsville Ship Canal. The rest of the beach is disappearing. Much of the developed area will be destroyed in a large storm. Ask some of the real estate people on the island about storm danger. You will probably not hear about Hurricane Beulah, which destroyed a number of buildings just a mile north of the jetty in 1967.

Access by Route 48 from Brownsville.

La Jolla, California. Just north of San Diego at La Jolla is a good place to see wave-cut terraces in the land that is slowly rising out of the sea as the earth moves and heaves. In the area of Mount Soledad the terraces rise up to eight hundred feet above the ocean. Between the jerky uplift of earthquakes, rising sea level cuts slowly into the cliffs. Leaping and diving from the La Jolla cliffs has been a favorite local sport, as is swimming and surfing on the long, sandy public beach. From the top of the cliffs just north of the Scripps Institute you can sometimes see the irregular wave patterns caused by deep submarine canyons cut into the continental shelf. Because there is less bottom drag, wave fronts advance more rapidly over the canyons.

Malibu, California. Here is a good example of the problems caused by development placed in a very dynamic give-

and-take between land and ocean. The strip of elbow-to-elbow beach houses is as poorly located as a baby sitting on the line of scrimmage at a football game. In the winter of 1978, mudslides from the hills below the coastal highway flowed into the houses from one side while twelve-foot waves battered them from the other. Almost all the sandy beach has disappeared and residents are trucking in boulders to protect the pilings of their homes. Scientists have observed much larger waves rolling into this same beach.

Bolinas Spit, California. This is a north-reaching sand spit, now developed with a private road and expensive houses. In the winter of 1978, storms cut into the spit as they do on barrier islands. Here is a typical West Coast example of unwise development. Homeowners immediately began building stone walls to hold their lots while petitioning the government for aid in building a seawall or revetment. A view down the row of beachfront houses will show you how quickly they all might disappear if major storm waves strike. Barrier spits like this are riskier than most barrier islands because they are usually thinner and not well vegetated.

Just off U. S. Route 1 a few miles above San Francisco and the Golden Gate Recreation Area.

Point Reyes National Seashore, California. Just in time to head off a subdivision, the federal government acquired these 64,000 acres for the West Coast's only national seashore. These granite hills covered with dense forests of fir and broad grasslands are a "geological island." The Point Reyes peninsula is thirty-five miles north of San Francisco, but was once far south. The point is separated from the rest of the continent by the San Andreas Fault, and the promontory has been creeping northward some two inches a year.

The park offers spectacular views of the coast and ocean from its 1,000-foot-high hills. A steel lighthouse perches on the tip of the headland and some 300 feet above the ocean. This area offers the best views of the waves wrapping themselves around the peninsula. These cliffs are also excellent places to view wildlife. Bird counters at Point Reyes have sometimes logged more than 190 species within a fifteen-mile radius of the point.

Try to go in the off season to avoid some of the point's 1.5-million annual visitors. Access by Route 1 from San Francisco.

Oregon Dunes National Recreation Area, Oregon. North of Coos Bay Route 101 runs along the back of what at times seem to be sandy hills covered with trees. These are fossil dunes, some over a hundred feet high, formed long ago on this flat section of the Oregon coast. Some have blocked small streams to form quiet ponds and lakes. Walking toward the ocean from one of these ponds, you enter the active dune fields. The entire dune field is almost fifty miles long, the longest coastal dune field in the United States. The higher dunes near the ocean offer many lookouts from which you can survey great lengths of beach and ocean. Many roads and paths off the highway provide access to a variety of public areas for picnicking and camping.

Newport, Devil's Punchbowl, Otter Crest, and Cape Foulweather, Oregon. Hundreds of headlands jut into the Pacific along the West Coast, but the area just north of Newport, Oregon, offers as much variety as anywhere we know. Newport's sandy beaches lie just in front of a sea terrace of fragile sandstone. Many houses have fallen in and become part of the ocean view they enjoyed. The sea seems to be claiming land at a rate of some two feet a year.

The hard lava of Yaquina Head juts almost a mile into the ocean and provides spectacular views of the coast, as if from a high vantage point at sea. Five miles north you can look down into the blue cove of Devil's Punchbowl spiked with pointed rocks. From the high south side you can get a good view of small sea terraces on the north. These were carved before the earth's crust lifted the rock formations out of the water.

A short distance north are Otter Crest and Cape Foulweather, two more hard-rock headlands. This is a rugged coast with crashing breakers and good lookouts for sea otters and birds.

This stretch of coast is reached by small roads leading off Route 101. Many picnic sites and small parks.

Cape Flattery, Olympic National Park, Washington. This is the most rugged beach scenery in the lower forty-eight states. In the most northwest corner of the Olympic peninsula, Cape Flattery looks out across the strait of San Juan de Fuca to Canada and the beginning of the glacial beaches. The cape itself is a 400-foot-high volcanic rock promontory. It is part of the Makah Indian Reservation.

Just east of the cape along the strait of San Juan de Fuca, the small town of Neah Bay offers travelers accommodations; and its relatively low-energy beach is a good place to look for driftwood and agates.

Those who prefer a good hike and solitude can hike south along the ocean beach for at least thirty miles. A favorite stopping place is Point of Arches, where huge rock formations separated from the mainland create fantastic shapes against the ocean horizon. At the foot of these "sea stacks" at ebb tide large clear tidal pools cradle many forms of sea life. In offshore islands visitors might see harbor seals resting.

Access is by Route 112 west from Port Angeles to Neah Bay, and from there on foot.

Bibliography of Useful References

Abbot, R. J. *American Seashells*. Princeton, N.J.: D. Van Nostrand, 1955.
The best single American shell book; beautifully illustrated.

Abbot, R. T. *Seashells of North America*. New York: Golden Press, 1968.
A pocket book useful for quick identification of common shells.

Barrier Islands Workshop. *Barrier Islands and Beaches*. Washington, D.C.: The Conservation Foundation, 1976.
A series of "lightly technical" papers on barrier island environments and resources and management approaches.

Bascom, W. *Waves and Beaches*. Garden City, N.Y.: Doubleday, 1964.
A readable description of the principles of waves and beaches; well illustrated and interestingly written.

Berrill, N. J., and Berrill, M. *The Life of Sea Islands*. New York: McGraw-Hill, 1969.
Concerned with the plants and animals of American barrier islands.

Bird, E. C. F. *Coasts*. Cambridge, Mass.: M.I.T. Press, 1968.
A well-illustrated treatment of different coastal types from a geological viewpoint, with emphasis on Great Britain.

Boesch, D. F.; Hershner, C. H.; and Milgram, J. *Oil Spills and the Marine Environment*. Cambridge, Mass.: Ballinger, 1974.

A summary of the suspected role of oil spills in ocean and beach ecology.

Brown, L. F., and others. *Natural Hazards of the Texas Coastal Zone*. Austin, Texas: Bureau of Economic Geology, 1974.
An absolute must for Texas beach dwellers and a highly useful book for anyone on an Atlantic or Gulf Coast beach. Discusses the hazards of coastal living and has very specific maps showing rates of shoreline erosion as well as other potential hazards.

Clancy, E. P. *The Tides*. Garden City, N.Y.: Anchor Books, 1969.
A readable discussion of the principles of ocean tides.

Clark, J. R. *Coastal Ecosystems: Ecological Considerations for Management of the Coastal Zone*. Washington, D.C.: The Conservation Foundation, 1974.
A useful volume for coastal zone planners, but somewhat detailed for individual homeowners.

Coastal Engineering Research Center. *Shore Protection Manual*. Washington, D.C.: U. S. Army Corps of Engineers, 1973, 3 vols.
The bible of shoreline engineering.

Coates, D. (ed.). *Coastal Geomorphology*. Binghamton, N.Y.: State University of New York, 1972.
A collection of technical papers on mechanisms of evolution and change of beaches and barrier islands.

Committee for Waterfront Structures (German). *Recommendations of the Committee for Waterfront Structures*. Berlin: Wilhelm, Ernst and John, 1971.
The bible of German ocean engineering.

Copeland, B. J., and Odum, H. T. "Ecological Systems by State," (pp. 104–23) in Odum, H. T.; Copeland, B. J.; and McMahan (eds.). *Coastal Ecological Systems of the United States*, Vol. I. Washington, D.C.: The Conservation Foundation, 1974.
Generalized sketch maps of coastal ecosystems from Maine through Texas.

Davies, J. L. *Geographical Variation in Coastal Development*. Edinburgh: Oliver and Boyd, 1972.
The topography and geomorphology of coasts.

Davis, R. A., and Ethington, R. L. (eds.). *Beach and Nearshore Sedimentation*. Tulsa, Okla.: Society of Economic Paleontologists and Mineralogists Special Publication #24, 1976.
Technical papers with recent developments in beach geology. A necessity for professional beach planners.

Davis, R. A. *Coastal Sedimentary Environments*. New York: Springer Verlag, 1978.
Excellent technical review papers on the processes of sediment movement in the major nearshore environments.

Dolan, R., and McCloy, J. *Selected Bibliography on Beach Features and Related Nearshore Processes*. Baton Rouge, La. Louisiana State University Press, 1965.

Ducsik, D. W. *Shoreline for the Public: A Handbook of Social and Legal Considerations Regarding Public Recreational Use of the Nation's Coastal Shoreline*. Cambridge, Mass.: M.I.T. Press, 1974.
One of the many useful volumes for the coastal zone manager to come from the M.I.T. sea grant program.

Georgia Department of Natural Resources. *Handbook: Building in the Coastal Environment*. Atlanta: State of Georgia, 1975.
A clearly illustrated, generalized summary of the principles of safe island-beach living and construction.

Hay, J., and Farb, P. *The Atlantic Shore: Human and Natural History from Long Island to Labrador*. New York: Harper & Row, 1966.
Shore plants and animals of New England.

Heppner, F. H., and Gould, L. L. "Birds of the Atlantic Seaboard from Cape Hatteras to Cape Cod" (pp. 8–1 to 8–128) in Saila, S. B., (ed.). *Coastal and Offshore Environmental Inventory: Cape Hatteras to Nantucket Shoals*. Providence: University of Rhode Island, Marine Publications Series No. 2, 1973.
An excellent guide to coastal and offshore birds.

Hoyt, J. H. *Field Guide to Beaches*. Falls Church, Va.: American Geological Institute, Earth Science Curriculum Project Pamphlet Series PS-7, 1971.
A primer on beaches, what they are made of and how they evolve.

Hunt, L., and Groves, D. (eds.). *A Glossary of Ocean Science and Undersea Technology Terms*. Arlington, Va.: Compass Publications, 1965.

Ingle, J. C. *The Movement of Beach Sand*. Amsterdam: Elsevier, 1966.
A classical study of beach sand movement on southern California beaches.

Ippen, A. J. *Estuary and Coastline Hydrodynamics*. New York: McGraw-Hill, 1966.
Shallow-water oceanography, tides, waves, and currents of the nearshore zone.

Johnson, D. W. *Shore Processes and Shoreline Development*. New York: John Wiley and Sons, 1919.
An old classic. The first book on shoreline processes. Many of the principles outlined are still valid.

King, C. A. M. *Beaches and Coasts*, 2nd ed., New York: St. Martins Press, 1972.

The standard textbook on the geology of beaches. Emphasizes European examples but reviews the American examples as well.

Leonard, J. N. *Atlantic Beaches.* New York: Time-Life Books, 1973.
A beautifully illustrated description of beaches from Cape Cod to Cape Lookout.

Manley, S., and Manley, R. *Beaches: Their Lives, Legends, and Love.* Philadelphia: Clinton, 1968.
A colorful account of the lore of beaches.

McGill, J. T. *Selected Bibliography of Coastal Geomorphology of the World.* Los Angeles: University of California at Los Angeles, 1960.

McHarg, I. L. *Design with Nature.* Garden City, N.Y.: Doubleday, 1971.
One chapter discusses design of a development on a typical barrier island.

Muir Wood, A. M. *Coastal Hydraulics.* London: Macmillan, 1969.
Nearshore oceanography and its application to shoreline engineering.

National Resources Defense Council, Inc. *Who's Minding the Shore? A Citizen's Guide to Coastal Zone Management.* Washington, D.C.: NOAA, Department of Commerce, Office of Coastal Zone Management, 1976.
A citizens' handbook on the Coastal Zone Management Act.

Pilkey, O. H. Jr.; Pilkey, O. H., Sr.; and Turner, R. *How to Live with an Island.* Research Triangle Park, N.C.: North Carolina Science and Technology Research Center, 1975.
A book concerned with Bogue Bank, North Carolina, with application to most barrier islands. Discusses hazards, shoreline erosion rates, construction techniques, and site choice.

Pilkey, O. H., Jr.; Neal, W. J.; and Pilkey, O. H., Sr. *From Currituck to Calabash.* Research Triangle Park, N.C.: North Carolina Science and Technology Research Center, 1978.
Similar to *How to Live with an Island* but covers the entire coast of North Carolina. Includes maps classifying the entire shore into unsafe, safe or cautionary in terms of development suitability.

Ricketts, E. F., and Calvin, J. *Between Pacific Tides,* 3rd ed. Stanford, Calif.: Stanford University Press, 1952.

Shepard, F. P., and Dill, R. F., *Submarine Canyons and Other Sea Valleys.* Chicago: Rand McNally, 1966.
A fascinating book summarizing what we know of canyons and their role (in some cases) in removing nearshore sand from the beach system.

Shepard, F. P., and Wanless, H. R. *Our Changing Coastlines*. New York: McGraw-Hill, 1971.

Shepard, F. P. *Submarine Geology* (3rd ed.). New York: Harper & Row, 1973.
The principles of geological oceanography applied to the world's oceans with an excellent chapter on beaches and coasts.

Steinhart, C. E., and Steinhart, J. S. *Blowout, A Case Study of the Santa Barbara Oil Spill*. North Scituate, Mass.: Duxbury Press, 1971.
An objective and readable history of the political, environmental, and economic factors involved in the oil spill that gummed up the beaches of Santa Barbara.

Sutton, A., and Sutton, M. *The Wild Shores of North America*. New York: Audubon Society, 1977.
Profusely illustrated documentation of our vanishing wild shorelines.

Sylvester, R. *Coastal Engineering*. Amsterdam: Elsevier, 1974, (2 vols.).
A detailed summary of the principles of ocean engineering.

Tricker, R. A. R. *Bores, Breakers, Waves and Wakes: An Introduction to the Study of Waves on Water*. New York: American Elsevier, 1964.
Textbook on the principles of all kinds of ocean waves.

U. S. Army Corps of Engineers. *Land Against the Sea*. Ft. Belvoir, Va.: Coastal Engineering Research Center, 1964.
An introduction to shoreline processes from the engineer's viewpoint.

U. S. Army Corps of Engineers. *The National Shoreline Study*. Washington, D.C.: U. S. Government Printing Office, 1973, (5 vols.).
These volumes represent the Corps' answer to the shoreline erosion problem of the United States. The entire U.S. shoreline is classified as to the need for engineering; $2 billion is suggested as being needed immediately.

U. S. Army Corps of Engineers. *Publications List, Coastal Engineering Research Center (CERC) and Beach Erosion Board (BEB)*. Ft. Belvoir, Va.: Coastal Engineering Research Center, 1976.
A listing of published research by the Corps on shoreline engineering. A must for beach scientists and planners.

U. S. Forest Service. *Houses Can Resist Hurricanes*. Madison, Wis.: U. S. Forest Service, 1965.
Well-illustrated booklet with construction details for hur-

ricane-resistant construction. Pole house construction is treated in particular detail.

Ursin, M. J. *Life in and Around the Salt Marshes.* New York: Thomas Y. Crowell, 1972.
A description of the plants and animals of the salt marshes.

Warmke, G. L., and Abbot, R. T. *Caribbean Seashells.* Narberth, Pa.: Livingstone, 1961.
A good one for the South Florida beachcomber.

Warner, L., and Strauss, D. *Inventory of the Barrier Islands of the Southeastern Coast.* New York: Open Space Institute, 1976.
An island-by-island rundown of ownership, development intensity, island natural characteristics, and probable future directions.

Weigel, R. L. *Oceanographical Engineering.* Englewood Cliffs, N.J.: Prentice Hall, 1964.
The principles of ocean engineering, including aspects of shoreline engineering.

Wlaton, T. L. *Hurricane-Resistant Construction for Houses.* Gainesville, Fla.: Florida Sea Grant Publication, University of Florida, 1976.

Zenkovitch. *Processes of Coastal Development.* Edinburgh: Oliver and Boyd, 1967.
Translated from the Russian. A technical summary of beach and island processes.

Index

Abilene, Kans., 51
Absecon Island, 169
Acad:a National Park, 294–95
"Air hammers," 33
Air rights, ownership of, 36
Alabama, 7, 19, 37
Alaska, 16, 25, 28, 30, 71, 270, 283; earthquake (1964), 118, 121, 127–28; glacial shoreline, 63–64; landslides, 119–22; tsunamis, 123, 124
Aleutian Islands, 123, 124
Alexandria, Egypt, 153
Alligator Alley (highway), 45
Altamaha Inlet, 102
Altamaha River, 38
Amazon River, 69–70
American Civil Liberties Union, 243
American Society of Civil Engineers, 188
Anaheim Bay, 82
Anchorage, Alaska, 72
Apalachicola River, 46
Appalachian Mountains, 28, 30–31
Arizona, 58
Ash Wednesday storm (1962), 69, 91–92, 95, 167, 299
Assateague Island, 202, 258, 261, 299
Astoria, Ore., 61
Aswan Dam, 58
Atchafalaya River, 49
Atlantic Beach, N.C., 82
Atlantic City, 91, 161, 162, 211, 257, 272, 282; development of, 169–72
Atlantic Ocean, 2, 19, 28, 32, 34, 35, 50, 91, 151, 188, 190, 212; hurricanes, 129; normal tide ranges, 72–73
Atlantis, 18
Atransas Pass, Tex., 73
Atrium Condominium Complex, 269
Augusta, Ga., 37
Australia, 115, 123
Avon, N.C., 258

Badlands, S.D., 28
Bahia Honda State Park, 302–3
Baltimore, Md., 245
Bank of the Islands, 268
Barcelona, port of, 153
Bar Harbor, 31
Barnegat Inlet, 82, 196
Barnstable, Mass., 161
Barrier islands, 24, 38, 45–46, 50, 51, 64, 94–113, 160, 175, 218, 273; and barrier splits, 112–13; birth of, 96–97; checklist for buying or building on, 286–87; dune systems, 109–13; effects of inlets, overwash, and dune movements on, 98–112; Florida, 107–8; Georgia sea islands, 97, 98, 102; Gulf Coast, 97, 99; New York, 106; North Carolina, 104–6, 108, 109, 112; sea level and, 97–98, 111; Texas coast, 100–2; Virginia coast, 94–97; West Coast, 97
Barton, Clara, 173
Bascomb, Willard, 80, 127
Bay City, Tex., 18
Bay Marchand, La., 278–79
Bay Ocean, Ore., 199–200
Bay of Fundy, 70–71
Bay of Naples, 153
Bay St. Louis, Miss., 73, 138–39, 141
Baytown, Tex., 21
Beaches: commercial development of

(in America), 164–87; how to live with, 252–84; number available for recreation, 224; rate of development, 227; seasonal changes in, 88–90. *See also* Dynamic equilibrium; Moving beaches; Sandy beaches
Beach property ownership, 223–51; California, 228–29, 234–35, 244, 249; cost of, 226; customary rights doctrine, 241, 249; English common law on, 231, 236, 241; Florida, 239–40, 249; Great Lakes area, 225–27; how to choose a house, 291–92; implied dedication principle, 237–38; low-tide line, 231–32; Oregon, 240–41; piers, 235; prescriptive easements, 239–40, 245–46, 247–48; public access question, 229–51; Roman law on, 230–31; Supreme Court on, 232–33; Texas, 240–41, 249; vegetation line, 241, 248
Beach residences, planning (for the future), 275–84; Bellah's approach to, 282–83; bridges, 280–81; buildings, 276–78; roads, 279–80; utilities, 278–79; where to build, 275–76
Beaufort, N.C., 10, 97, 135–36
Bella, David, 221, 282–83
Berm, 89, 92
Bermuda Triangle, 79
Big Shell Beach, 305
Biscayne Bay, 133, 176–79, 185
Blind Pass, 107
Bogue Bank, 9–10, 104–6, 144
Bolinas Spit, Cal., 307
Boots, Edmund, 206
Bores, tidal, 69–70
Boston, Mass., 18, 31, 72, 245
Boyd, Justice, 239–40
Brazos River, 51
Breakwaters, 128, 189, 192, 220; engineered shorelines and, 203–7
Bridgeport, Conn., 73
Broward County, Fla., 225
Brownsville Ship Canal, 306
Brunn's Rule, 25
Buchanan, James, 165
Budd, Thomas, 161
Bulger, William, 243
Bulkheads, 192, 209, 220, 225, 253, 261, 263–64, 270; engineered shorelines and, 207–8
Bunker Hill, 33
Buxton, N.C., 214–15, 258, 260

Cabot, John, 160
Calcium carbonate, 41
California, 16, 86, 91, 93, 127, 204–5, 244, 249, 265–67, 306–8; beach property ownership, 228–29, 234–35, 244, 249; cliffs and bluffs, 58–60, 115; disappearing beaches (in the south), 51–58; earthquakes, 52; landslides, 116–19; piers in, 86–88; river sediment, 56, 57, 61; storm waves, 130
California Supreme Court, 234–35, 237
Canada, 19, 31, 32, 63
Canute, King, 9, 222
Cape Canaveral, 70
Cape Cod, 13, 30, 32, 34–36, 56, 91, 161, 162, 264–65, 279; bluffs of, 35; February storm (1978), 35
Cape Cod National Seashore, 296
Cape Fear, 73
Cape Fear Inlet, 102
Cape Fear River, 38
Cape Flattery, 63, 309–10
Cape Florida Lighthouse, 196–97
Cape Foulweather, 308–9
Cape Hatteras, 12, 70, 73, 78–79, 91, 161, 184, 265
Cape Hatteras Lighthouse, 215, 258, 260
Cape Hatteras National Seashore, 213, 257, 258
Cape Kennedy, 73
Cape Lookout, 73, 97, 104, 110, 218, 257
Cape Lookout National Seashore, 245, 300
Cape May, 16, 191, 196, 211, 271; development of, 164–69
Cape Meares, 198
Cape Mendocino, 58–59, 60
Cape Romain, 73
Cape Romano, 45, 46
Caribbean hurricanes, 129
Carmel, Cal., 89
Carson, Rachel, 45, 68, 267
Carter, Jimmy, 97
Carthage, port of, 155, 156, 185
Cascade Mountains, 27–28
Catastrophes. *See* Natural disasters
Catskill Mountains, 17
Cedar Key, Fla., hurricane (1842), 149
Channel dredging, 107
Channel Islands, 85, 87
Chattahoochee River, 46
"Chenier" beaches, 42, 49
Chenier Caminada, 99; hurricane (1893), 150
Chesapeake Bay, 25, 70
Chevron Oil Company, 278–79
China, 69, 88
Civilian Conservation Corps (CCC), 213
Claudius, Emperor, 154, 155, 158
Cliffs and bluffs: in California, 58–60, 115; checklist for buying or building on, 287
Cliffs and coves, battered shores of, 58–63
Coastal Area Management Act, 10
Coastal Energy Impact Funds, 210
Coastal Zone Management Act (1972), 246–48
Coastal Zone Management Agency, 36
Cobble beaches, slope of, 90
Cohen, Hal, 181
Collins, John, 175–76, 177
Colorado River, 58, 305
Columbia River, 62–63, 72
Commercial development, 164–87;

Atlantic City, 169–72; Cape May, 164–69; coastal management and, 183–87; Miami Beach, 174–83, 185; Yankee land speculators (post-Civil War), 172–73. *See also* Engineered shorelines
Coney Island, 272, 297–98
Connecticut, 5, 40; years likely between storms, 135
Connecticut Civil Liberties Union, 244
Conservation Council of North Carolina, 6, 8–9
Conservation Foundation of North Carolina, 106
Continental shelf, 39, 40, 86, 96; sand deposits on, 38; underwater mapping of, 87–88; waves and, 70–71
Coos Bay, 61–62, 72
Corals, 42, 44
Core Banks, 110, 218
Corpus Christi hurricane (1919), 150
Crescent City, Cal., 61, 72, 128
Crown Zellerbach Corporation, 211–12
Crustaceans, 45
Cuba, 139
Cumberland Island, 97
Curicant Dam, 58
Currents, 19, 71, 81, 82; longshore (or littoral), 81–83, 86, 192; sand flow due to, 82
Currituck Banks, 94–97
Currituck Sound, 94, 95
Customary rights, doctrine of, 241, 249
Cyclones, 115

Dade County, Fla., 197
Dallas, Tex., 51
Dams, 220
Davis, Jefferson, 143
Daytona Beach, Fla. 73, 239
Deal, N.J., 246
Death Valley, 151
Delaware, 231–32, 262–63, 299; years likely between storms, 135
Delaware Bay, 164
Delaware River, 38
Delft, Netherlands, 250
Del Mar, Cal., 53, 54
Del Mar Beach Club, 55
Delta Project (Holland), 158
Deltona Corporation, 45
Deroucher Construction Company, 197
Devil's Punchbowl, Ore., 308–9
Diamond Head (Hawaii), 84–85
Dietz *v.* King, 237, 238
Dike system (Netherlands), 156–60
Dolan, Dr. Robert, 213, 258
Doughty, Gen. Enoch, 169–70
Driftwood, 30, 44, 63
Drumlins, 33, 34
Dry Dock Company, 181–82
Duane, Dr. David, 279
Duke Marine Laboratory, 8, 9
Duke University, 6, 7, 105
Dynamic equilibrium, 15–20, 191, 218, 270; barrier islands, 93, 94–113; disasters and, 114–50; energy (wind, waves, and tides), 15, 65–83, 84–93, 131; materials (of the beach), 15, 27–64, 93, 131; sea level, 15, 17–26, 39, 41, 42–45, 52, 85, 97–98, 111, 131, 158, 183, 220; shape of the beach, 15, 84–93, 131; *See also* Moving beaches

Earthquakes, 52, 67, 118; and landslides, 118
East Coast: beach materials (from New York to Florida), 37–42; checklist for buying or building on, 285–92; size of beach and sand available, 90; swells, 93
East Coast Railroad, 43
Eastham, Mass., 161
Eastport, Me., 72
East River, 71
Eastward (research vessel), 8
Ebb tidal deltas, 103
Eckhardt, Robert, 242
Edge, Billy, 217
Ediz Hook, 211–12
Egypt, 58, 151
Elliot Key, 133
Emerald Isle (resort development), 104–6
Encinitas, Cal., 51, 55, 266; disappearance of streets in (1938), 52–53
Energy (winds, waves, and tides), 15, 65–83, 84–93, 131; forms of waves, 75–80; longshore (or littoral) currents, 81–83; tides, 68–71; wave, 65–67; wind and, 73–75
Engineered shorelines, 188–222; break-waters, 203–7; consequences of, 191–92, 218–22; differences over, 219; goal of, 192; groins and jetties, 193–203; headland system, 189; replenishment and dune building, 212–18; seawalls, bulkheads, and revetments, 207–12. *See also* Commercial development
English common law, 231, 236, 241
Environmental Protection Agency (EPA), 256
Ervin, Justice Richard W., 240
Etruscans (ancient), 151
Eureka, Cal., 59, 61
Everglades, 43, 64, 179
Everglades National Park, 45

Fairfield, Conn., 244
Farmers Home Administration, 256
Federal Writer's Project (1938), 136–37
Feldspar, 39–40
Field, Elnathan, 175
Field, Justice Stephen J., 232
Fifth Amendment, 269
Fiords, glacial, 64
Fire Island National Seashore, 106, 257–58, 261–62, 297
Fisher, Carl, 176–79, 180

Fitzgerald, F. Scott, 36–37
Flamingo Gorge Dam, 58
Flint River, 4
Flood walls, 21
Florida, 25, 28, 37, 70, 97, 102, 109, 162, 206, 225, 235, 249, 267–69, 270, 275, 302–3; barrier islands, 107–8; beach property ownership, 239–40, 249; hurricanes, 133–34, 149, 150; river sediments, 45–47; story of Miami Beach, 174–83, 185; trees against the sea, 42–45; years likely between storms, 135
Florida Bay, 43
Florida Keys, 41, 42–43, 133–34, 149, 272
Florida Supreme Court, 239–40
Floyd Bennett military airfield, 263–64
Ford, Henry, 166
Fort Macon, 10
Fort Myers, Fla., 43
Fourteenth Amendment, 244
Frank, Dr. Neil, 135, 149
Frog Hollow, N.J., 167–68
Frost, Robert, 80

Galt Ocean Mile (Florida), 225
Galveston, Tex., 21, 50, 73, 82, 202, 238, 304–5
Galveston Bay, 21, 202
Galveston County, Tex., 18
Galveston hurricane of 1900, 9, 15, 150, 185, 209
Garvey, Frank, 167
Gateway National Recreation Area, 263
Georges Bank, 31, 34
Georgia, 70, 91, 109, 172, 232, 301–2; years likely between storms, 135
Gewirtz *v.* City of Long Beach, 243
Gion *v.* City of Santa Cruz, 237
Glaciers, 33, 34, 38, 42, 52; in Alaska, 63–64
Glen Canyon Dam, 58
Godfrey, Dr. Paul, 96, 214, 259, 264
Goldsmith, Dr. Victor, 182
Goss, Porter, 268
Grand Bank, 31
Grand Canyon, 17, 28
Grand Cayman Island, 139
Grand Isle, La., 30, 99–100, 304; hurricane (1831), 149
Grant, Ulysses S., 165
Gravel beaches, 93
Gray's Harbor, 63
Great Lakes, 225–27
Great Lakes Dredge Company, 181–82
Great Plains, 38
Great South Bay, 297
Great Wall (cliff-top condominium), 54
Great Wass, Island, 294
Greece (ancient), 123, 151, 152–53, 163
Green Hill Beach, 296–97
Greenhouse effect, 18
Green Pond, R.I., 212
Greenwich, Conn., 246
Groin system, 192, 193–95, 205, 209, 220, 235–36, 261, 270
Gross National Product, 162
Grotius, Hugo, 250–51
Gulf Coast, 7, 16, 43, 45, 46, 61, 267; checklist for buying or building on, 285–92. *See also* names of states
Gulf Islands National Seashore, 303–4
Gulf of Alaska, 79
Gulf of California, 58
Gulf of Maine, 31, 71
Gulf of Mexico, 20, 28, 30, 47, 50, 99, 101, 202, 210, 257, 267; normal tide ranges, 73
Gulf Oil Company, 47, 278–79
Gulfport, Miss., 138–39

Haining, China, 69
Hale, Lord, 233
Hancock County, Miss., 145, 256
Harker's Island, 97
Harris, Joel Chandler, 173
Harrison, Benjamin, 165
Hatteras Island, 112, 171
Hawaiian tsunamis, 123–27
Hay, William and Georgianna, 240–41
Hayes, Dr. Miles, 144, 301
Headland system, 189, 217
Hebrews (ancient), 151
Hell Gate, 71
Hermosa Beach, 204
Hicks, Dr. Stacey, 19, 21
Higgins Beach, 295
Hillsboro Inlet, 82
Hilo, Hawaii, 123–27
Hiltonhead Island, 172, 173; storm surge floods (1893), 150
Ho, Siew-Koon, 188–89, 217
Holly Beach, 49
Homesites, how to choose, 288–91
Honolulu, Hawaii, 72
Hoover Dam, 58
Houston, Tex., 21, 50, 245
Hudson, Henrik, 160
Hueneme Submarine Canyon, 197
Hughes, Ellis, 164–65
Humble Oil Refinery, 21
Hunting Island, 301
Hurley, W. L., 173
Hurricane Agnes, 134
Hurricane Betsy, 99, 139, 141, 142, 272
Hurricane Beulah, 306
Hurricane Camille, 7, 8, 116, 134, 138–42, 228, 272, 304; casualties and destruction, 150; lessons of, 144–49
Hurricane Carla, 144
Hurricane Carol, 150
Hurricane Ella, 256
Hurricane Hazel, 144
Hurricanes, 128–32, 178–79, 272; Atlantic and Gulf Coast, 134–50; as "disturbances," 129; Federal Writer's Project report on (1938), 136–37; list of, 149–50; property ownership and,

132–42; rating of, 134; recorded casualties, 131; shoreline equilibrium and, 131–32; wind velocity and energy from, 129–30. *See also* Storms; names of hurricanes

Icebergs, 64
Icecaps, 18, 20, 32
Iceland, 20
Illinois, 232
Illinois Central Railroad Company, 232
Imperial Dam, 58
Implied dedication, principle of, 237–38
Indian Ocean, 115
Indianola hurricane (1886), 149
Indian River Delaware Inlet, 73
Indians, 4, 161, 171, 238
Inlets, 171, 201, 213, 261; filling of, 107; flood deltas of, 103–4; interdependency of, 103. *See also* names of inlets
Internal Revenue Service (IRS), 106
Irish Cove Point, 60
Isaacs, John, 205
Isle Derniere, 210

Jackson, Henry, 243
Jackson Gap, 200
Jamaica Bay, 40, 263–64
Japan, tsunami of 1896, 123
Japan Current, 19
Jekyll Island, 73, 97
Jetties, 192, 195–203, 205, 220, 270, 306
J. N. Darling Wildlife Refuge, 267
Jockey's Ridge, 112
John, King, 231
John Pennekamp State Park, 43
Johnson, Andrew, 172
Jonesport, Me., 294
Juneau, Alaska, 72, 119

Kennedy, John F., 296
Kennedy Airport, 40, 263–64
Ketchikan, Alaska, 72
Key Biscayne, 182, 196, 212
Key Largo, 275
Key West, 73, 165–66; hurricane (1846), 133, 149; hurricane (1919), 150
Kiawah Island, 300–1
Kitty Hawk, N.C., 258
Klem, F. R., 173
Komar, Dr. Paul, 201–2
Kuhn, Gerry, 55

Labor Day storm of 1935, 150
Labrador Current, 78
Lafitte, Jean, 47
Laguna Dam, 58
Laguna Madre, 101–2, 305
La Jolla, Cal., 55, 72, 89, 90, 210–11, 265–66, 306
Lake Borgne, 46
Lake Erie, 28–29
Lake Pontchartrain, 46
Landslides, 115, 116–22; in Alaska, 119–22; conditions conducive to, 53–54; earthquakes and, 118; Pacific shore, 116–19
Lee, Robert E., 10
Lichens, 33
Lighthouses, 80, 166
Limestone, 44, 46
Lincoln, Abraham, 165
Lincoln City, Ore., 62
Little Shell Beach, 305
Lituya Bay, 119–20, 122
Long Beach, Cal., 87–88, 91, 138–39, 141–42
Long Island, 1–5, 12, 25, 36–37, 69, 161, 171, 179, 186, 209, 228, 243, 272; bluffs of, 35; hurricane (1815), 149; summer storm (1938), 150
Long Island Sound, 2, 37, 71, 297
Longshore (or littoral) currents, 81–83, 86, 192
Los Angeles, 57, 72, 162, 204, 245
Louisiana, 30, 37, 42, 46, 48, 210, 278–79, 304; hurricane (1909), 150; river sediments, 47–50
Lum, Charles and Henry, 174–75

McCaffrey, Cheryl, 264
MacDonald, John, 107–8, 303
Magna Carta, 231
Magoon, Orville T., 189–90
Maine, 30, 32, 42, 43, 231–32, 270, 294–96; years likely between storms, 135
Malibu, Cal., 56, 130, 204, 271, 272–73, 306–7
Mangrove trees, rising sea level and (in Florida), 42–45
Manomet Bird Observatory, 36
Manomet Bluffs, 35–36
Manteo Island, 161
March storm (1962), 150
Marco Island, 45
Marconi, Guglielmo, 35, 296
Mare Liberum (Grotius), 250
Marine life, 41–42, 219
Marks (developer), 234–35
Marseilles, port of, 153
Martha's Vineyard, 34, 165
Martin *v.* Waddell, 233
Maryland, 109, 202, 261, 298–99; years likely between storms, 135
Masonboro Inlet, 196
Massachusetts, 19, 33, 162, 231–32, 242–43, 264–65, 279, 296; years likely between storms, 135
Matagorda Peninsula, 144
Materials (of the beach), 15, 27–64, 93, 131; Alaskan coast, 63–64; cliffs and coves (California), 58–63; East Coast, 37–42; Florida mangroves, 42–45; grain size and, 29; kinds of, 27–28; Long Island, 36–37; Northeast coastline, 30–36; Northwest coastline, 60–63; recipes for, 28–29; river sediments and

sea, 45–51; sand dunes, 61–63; sandy beaches, 36–42; seasonal effects, 29–30; southern California, 51–58
Mathewson, Dr. Christopher, 253–54
Mauna Loa volcano, 124
Mayport, Fla., 73
Mediterranean Sea, 151–53
Mendocino, Cal., 118–19
Mendocino County, Cal., 238
Meuse River, 159
Mexico, 37, 42, 58
Mey, Henrik, 160
Miami, Fla., 16, 41, 43, 148, 245; hurricane (1926), 150
Miami Beach, 42, 193, 195, 209, 211, 228, 268, 271, 272, 282, 293, 302; development of, 174–83, 185
Miami Harbor, 73; jetties at, 196–97
Mid-Atlantic Ridge, 20
Minnesota, 28, 46
Minshall, William, 226
Mission Beach, 56
Mississippi, 7, 37, 116, 138–39, 303–4; hurricane (1909), 150
Mississippi delta, 20, 46–48, 279
Mississippi River, 28, 46, 48, 99, 186, 210
Missouri River, 49
Mjerda River, 155
Mobile, Ala., 73, 245
Monmouth Beach, 168, 208–9, 298
Monroe, Commodore Ralph, 178
Montauk Point, 179
Moore, George, 55
Morehead, N.C., 97
Morgan, J. P., 2
Morgan's jetty, 2, 4, 6
Moriches Inlet, 106, 261, 262
Morton, Robert, 197
Mount Desert Isle, 31
Mount Rushmore, 64
Moving beaches: buying or building on, 285–92; commercial development and, 164–87; efforts to live with (in history), 151–63; future planning and, 275–84; how to live with, 252–84; nature of, 12–16; ownership of, 223–51; where to see, 293–310; *See also* Dynamic equilibrium; names of beaches
Mudflats, 44, 47
Murphy's Law, 127
Muscle Beach, 56
Mustang Islands, 50
Myrtle Beach, S.C., 73

Nag's Head, N.C., 23, 171, 203, 258
Nantucket, 34, 165
Naples, Fla., 45
Naples, Italy, 153
Nassau-Suffolk Regional Planning Commission, 217
National Flood Insurance Program (NFIP), 255–56
National Guard, 149
National Hurricane Center (Miami), 134, 135, 147, 149
National Marine Fisheries Building, 55
National Oceanographic Survey, 19
National Park Service, 214, 245, 257–66, 279–80, 297
National Shoreline Erosion Demonstration Program, 217
National Shoreline Study, 9, 11, 254
Natural disasters, 114–50; hurricanes and storms, 128–50; landslides, 116–22; tsunamis, 122–28
Nature Conservancy, 294
Nauset Beach, 296
Navajo Dam, 58
Neptune City *v.* Avon-by-the-Sea, 243
Netherlands, 162, 163, 185, 190; dike system, 156–60; major floods in, 159; storm surge of 1953, 142
Netherlands Waterways Authority, 156–57
Neumann, Dr. Conrad, 182, 183
Nevada, 38
New Currituck Inlet, 94
New England, 4, 22, 25, 28, 37, 39, 48, 69, 91, 131, 209, 221, 264–65, 297; August hurricane (1635), 149; coastline, 30–36; jagged nature of, 32–33; northeasterlies, 74–75; rising sea level, 19; shaping of, 31–32; summer storm (1938), 150; "sunken" coastline, 33–34. *See also* names of beaches; states
Newfoundland, 31, 32
New Hampshire, 32, 231–32; years likely between storms, 135
New Jersey, 5, 9, 16, 28, 69, 91, 109, 161, 176, 193, 196, 208–9, 232, 244, 272, 298; years likely between storms, 135
New Jersey Supreme Court, 243
New London, Conn., 73
New Mexico, 51
New Orleans, 46, 49, 50, 99, 186, 245
Newport, Ore., 61, 308–9
Newport, R.I., 72
Newport Beach, Cal., 86, 194–95
New York, 17, 18, 297–98; years likely between storms, 135
New York City, 40, 73, 245; sea level trends, 22
New York Civil Liberties Union, 244
Nile delta, 58
Nixon, Richard, 196
Norfolk, Va., 245
North Carolina, 6–10, 23, 41, 46, 70, 97, 102, 104–6, 108, 109, 144, 145, 171, 184, 186, 188, 196, 202–3, 245, 256, 274, 300; barrier islands, 104–6, 108, 109, 112; years likely between storms, 135
North Carolina State University, 6–7
North Carolina Supreme Court, 235
Northeast coastline, 30–36. *See also* names of beaches; states
North Sea storm (1953), 156–57, 185
Northwest coast, 60–63. *See also* names of beaches; states

Nova Scotia, 70–71

Oahu Island, 84–85
Ocean City, Md., 73, 82, 202, 271, 298–99
Ocean City, N.J., 176
Ocean City Inlet, 202
Ohio, 46
Old Orchard Beach, 295–96
Olympic National Park, 309–10
Open Beaches Act (Tex.), 242, 243
Open Shorelines Act (Virgin Islands), 242
Oregon, 16, 91, 127, 198–201, 308–9; beach property ownership, 240–41; coastal dunes, 61–63
Oregon Dunes National Recreation Area, 308
Oregon Highway Department, 240
Oregon Inlet, 202–3
Osborn, Ezra, 175
Ossabaw Island, 97
Ostia, port of, 154, 155, 254
Otter Crest, Ore., 308–9
Outer Banks (N.C.), 8, 102, 104, 145, 188, 202–3, 213, 216–18, 264, 274
Outlaw, E. R., 23
Oxnard, Cal., 82

Pacific Ocean, 19, 28, 39, 86, 115, 200, 257; normal tide ranges, 72; southern California, 51–58
Pacific Palisades, 117–18
Pacific sea cliffs, 25
Pacific Tsunami Warning System (PTWS), 127
Padre Island, Tex., 50–51, 100, 253
Padre Island National Seashore, 305–6
Palm Beach, Fla., 43
Palos Verdes, Cal., 117
Palos Verdes Hills, Cal., 117, 204
Pamet River, 34
Pamlico Sound, 203, 214
Panama Canal, 19
Parker Dam, 58
Parker Toys Company, 170
Pass Christian, Miss., 138–40, 141, 304
Pebble beaches, slope of, 90
Pennsylvania, 231–32
Penny's Hill, Va., 95
Pensacola, Fla., 73
Perdido Pass, Ala., 82
Perigean spring tide, 69
Perigee, moon, 68, 69
Petersburg, Alaska, 72
Petit Bois Island, 303–4
Philadelphia, 37, 245
Phillips Oil Company, 278–79
Phoenicians (ancient), 151, 155
Piedmont fall line, 38, 39, 61
Pierce, Franklin, 165
Pierce, Dr. Jack, 8, 214
Piers: in California, 86–88; public access to, 235
Piper, Donald, 253, 254
Piraeus, port of, 153
Pitney, Dr. Jonathan, 169–70
Plankton, 41
Pleistocene islands, 98
Plunging breakers, 76–77
Plymouth, Mass., 35–36, 161
Plymouth Rock, 31
"Pocket beaches," 42
Point Arena, 60
Point Lobos, 86
Point Reyes National Seashore, 307–8
Pollution, 18, 42, 219
Port Angeles, Wash., 211
Port Hueneme, 197
Portland, Me., 31, 72
Portland, Ore., 245
Port Leon hurricane (1843), 149
Port St. Joe hurricane (1844), 149
Port Townsend, Wash., 72
Portuguese Bend, Cal., 117
Prescriptive easement, principle of, 239–40, 245–46, 247–48
Presque Isle peninsula, 28–29
Private beaches, demand for, 224
Private property, law of, 223. *See also* Beach property ownership
Proposition 20 (Cal.), 244
Providence, R.I., 245
Provincetown, Mass., 34, 91, 264, 296
Public trust doctrine, 231
Puerto Rico, 40, 41

Quartz, 39–40, 182

Rachel Carson Seacoast, 294
Rainey, Roy, 173
Raleigh, N.C., 37
Raleigh, Sir Walter, 161, 171
Ramapo (freighter), 67
Red Cross, 149, 173
Redemption Act, 172–73
Redondo Beach, 86, 204–5
Reefs, 41
Replenishment, dune building, 212–18
Res communes, 230–31
Resurrection Bay, 121
Revetments, 192–93, 261; engineered shorelines and, 210–12
Rhine River, 159
Rhode Island, 30, 150, 212, 272, 296–97; bluffs of, 35; years likely between storms, 135
Rhodes, harbor of, 153
Richelieu Apartments party, 140
Richmond, Va., 37
Riggs, Dr. Stan, 203
Rio Grande River, 51, 306
Riomar Reef, 206
Rivers and Harbors Act, 184
River sediments, 45–51; in California, 56, 57, 61; islands off Texas, 50–51; Mississippi delta, 47–50; offshore Florida, 45–47

Roanoke Colony, 161
Rock Crusher Gap, 200
Rogers, Will, 178, 204
Roman law, 230–31
Rome (ancient), 151, 153–56, 163
Rome, Italy, 190
Roosevelt, Theodore, 9, 10
Rosen, Dr. Peter, 25
Runnels (or troughs), 90–91
Rusnak, Dr. Gene, 42
Rye, N.Y., 244

Sabecon Reef Association, 206
Sabine River, 50
Sacramento, Cal., 72
Sager, Lawrence, 243–44
St. Johns, New Brunswick, 72
St. Petersburg, Fla., 73, 107, 267, 271, 303
St. Simons Island, 97
Salvation Army, 149
San Andreas Fault, 60, 307–8
Sand: calcium carbonate part of, 41; mineral content of, 39–40; nature of, 24. *See also* Sandy beaches
Sandbars, 44
Sand dunes: barrier islands, 109–13; Oregon and Washington coast, 61–63
San Diego, Cal., 51, 54, 56, 118–19
San Diego Association of Geologists, 55
San Diego Quarantine Station, 72
Sand mining, 40
Sandstones, 52
Sandwich, Mass., 161
Sandy beaches, 33, 36–42, 56, 93, 175; East Coast (from New York to Florida), 37–42; environmental threat, 40; Long Island, 36–37; marine life, 41–42; mineral content, 39–40; sand deposits, 38–39
Sandy Hook, N.J., 73, 82
San Francisco, 18, 58, 59, 71, 72, 238, 245; 1906 earthquake, 118
Sanibel Island, 45, 267–69
Sanibel Planning Commission, 268
San Jacinto Battleground Park, 21
San Juan, P.R., 73
San Juan de Fuca, straits of, 63, 211, 309
San Pedro, Cal., 87
Santa Barbara, Cal., 72, 82, 118
Santa Clara River, 57
Santa Monica, Cal., 56, 77, 82, 117, 197
Santa Monica Bay, 204
Santa Monica Mountains, 85
Sapelo Island, 301–2
Sarasota, Fla., 267
Sargent, Tex., 305
Savannah hurricane (1893), 150
Savannah River, 38, 73
Scripps Institution of Oceanography, 54, 55, 205
Scripps Submarine Canyon, 56
Sea Around Us, The (Carson), 267
Sea Cliff, N.Y., 1–6
Sea Gull, Va., 95
Sea islands, 97, 172, 173
Sea level, 15, 17–26, 39, 41, 42, 43, 44, 52, 85, 97, 131, 158, 183, 220; barrier islands and, 97–98, 111; dune system and, 23–24; estimated losses, 22; long-term changes, 20; mangroves (in Florida) and, 42–45; measurement of, 19–20, 25; mid-ocean ridge factor, 20; in New York City, 22; shoreline erosion, 24–25; in Texas, 21
Seaside Park, N.J., 176
Seasonal changes, beach material and, 29–30
Seattle, Wash., 72, 245
Seawalls, 190–91, 192, 220, 225, 235–36, 249, 253, 261, 263–64, 270; engineered shorelines and, 208–10
Seaway Co. *v.* Attorney General, 238, 239
Seaweed, 44
Seismic waves, 122–23
Self-Realization Fellowship Building, 53
Seward, Alaska, 121
Seymour Narrows, 71
Shackleford Bank, 109, 300
Shackleford Island, 218, 245
Shadeberg, Henry, 226
Shale layers, 21, 52
Shallow water wave, 76
Shape of the beach, 15, 84–93, 131; horizontal changes, 91–92; seafloor and, 84–88; seasonal changes, 88–91
Shell Oil Company, 278–79
Shells, 41, 44
Shepard, Francis, 124–26, 265–66
Sheridan, Francis, 50
Sherman, Gen. William T., 172
Shingle beaches, 93
Shinnecock Inlet, 261
Ship Shoal Light, La., 73
Shively *v.* Bowlby, 233
Shoaling (fine-grained) beaches, 152
Shoals, 33, 78, 103, 201
Shoreface, 85
Shoreline, checklist for buying or building on, 287–88
Shoreline Management Act (Washington), 244–45
Shreveport (transport), 78
Shrimping industry, 58
Sierra Club, 293
Silent Spring (Carson), 45, 267
Siletz spit, 62
Siltstones, 52
Sloping (coarse-grained) beaches, 152
Smith, Lester, 36
Solana Beach, 55
Southampton, N.Y., 166
South Bethany, Del., 299
South Cape May, 167
South Carolina, 79–80, 97, 109, 172, 220–21, 256, 300–1; years likely between storms, 135
South Dakota, 28
Southern California, 51–58; earthquakes,

52; landslides and cliff erosion, 53–54, 56; 1975 study, 56–57; river sediments, 56, 57; sand-pumping operations, 58; sea level, 52; shoreline retreat, 55; water supply, 55–56, 57. *See also* California
Southern California Testing Laboratory, 55
South Padre Island, 101, 306
Spanish Head, 74
Spender, Stephen, 77–78
Sphaeroma (crustacean), 45
Spring tides, 68–69
Stevenson, Adlai, III, 226–27
Stick, David, 102
Storms, 49, 53, 54, 69, 79, 89, 93, 128–32, 276–77; air pressure, 130–31; mythology and religious views of, 138; property ownership and, 132–42; at sea, 129; shoreline equilibrium and, 131–32; wind velocity and energy from, 129–30. *See also* Hurricanes
Storm waves, 74–75, 79–80, 130, 200
Straits of Gibraltar, 151
Sullivan's Island, 79–80
Surfers, 76–77, 84–85, 91
Surtsey Island, 20
Susquehanna River, 38
Swells, 93
Sylvester, Richard, 188–89, 217
Syzygy cycle, 68–69

T. B. Potter Realty Company, 198–99
Taft, Ore., 74
Tamiami Trail, 45
Tampa, Fla., 45, 46, 245, 303; hurricane (1848), 149
Taney, Roger, 233
Taoism, 88
Terich, Thomas, 201
Terpen (mounds of earth), 159
Texaco Oklahoma (tanker), 78–79
Texas, 22, 25, 37, 42, 48, 50–51, 100, 144, 186, 197, 202, 240–41, 243, 249, 304–5; beach property ownership, 240–41, 249; hurricane (1919), 150; river sediments, 50–51; southeasterlies, 74; underground shale layers, 21
Texas City, 21
Texas Geological Survey, 197
Thornton *v.* Hay, 240–41
Tiber River, 153, 154
Tides, 2, 3, 13, 19, 68–71; cresting or subsiding wave, 71; and currents, 71, 81, 82; height or "range" of, 70; influence of sun and moon on, 68–69; normal ranges, 72–73; spring, 68–69. *See also* Waves and wave energy
Tillamook Bay, Ore., 72, 200
Tillamook Inlet, 198
Tillamook Rock Lighthouse, 80
Tillamook Spit, Ore., 16, 199
Tomales Bay, Cal., 234
Tombigbee Inlet, 102
Tona-Rama, Inc., 239
Tongass Forest, 28
Trade-offs, environmental, 191
Trajan, Emperor, 154, 155, 158
Trans-Alaska pipeline, 120–21
Treasure Island, Fla., 107, 303
Trieste, port of, 153
Truro, Mass., 34, 91
Tsientang River, 69
Tsunamis, 71, 121, 122–32; in Alaska, 123, 124; first recorded, 123; Hawaii, 123–27; ideal conditions for, 122–23; in Japan, 123; speed of, 122, 123; warning systems, 126–27; West Coast, 127–28
Turkey Point, Fla., 182–83
Typhoons, 115

Udall, Stewart, 94
Underwater bars, 38
U. S. Army Corps of Engineers, 9, 11, 24, 28–29, 34, 36, 40, 45, 47, 49, 57–58, 82, 100, 106, 130, 141, 144, 167, 168, 171, 178, 181, 184, 190, 194, 196, 197, 198, 199, 200, 201, 202–3, 205, 209, 210, 211, 214–15, 226, 227, 228, 246, 254–55, 257–58, 260, 261–62, 279, 301, 304
U. S. Constitution, 244, 269
U. S. Department of Housing and Urban Development, 167, 168
U. S. Fish and Wildlife Service, 294
U. S. Geological Survey, 12, 42, 142
U. S. Supreme Court, 232–33
U. S. Weather Bureau, 34
University of Delaware, 262–63
University of Georgia Marine Institute, 302
University of North Carolina, 6
University of Texas in Austin, 253
Utah, 38
Utica, port of, 155

Valdez fiord, 120–21
Vallianos, Lim, 190–91, 228
Vancouver, B.C., Canada, 72
Vancouver, Wash., 72
Vegetation line, 241, 248
Ventura River, 57
Verrazano, Giovanni da, 160
Vespucci, Amerigo, 12
Victoria, B.C., Canada, 72
Vinellas County, Fla., 82
Virginia, 41, 88–89, 109, 231–32, 261; barrier islands, 94–97; years likely between storms, 135
Virginia Beach, 23, 73, 88–89, 94, 271
Virginia Coast Reserve, 299–300
Virginia Institute of Marine Sciences, 25, 182
Virgin Islands, 40, 242

Waco, Texas, 51
Waikiki Beach, 77, 82, 84–85
Walker, Ronald, 258, 259–60

Wallace, McHarg, Roberts, and Todd (planning firm), 268
Ward's Island, 71
Washington, 211–12, 244–45, 309–10; coastal dunes, 61–63
Washington, D.C., 37, 245
Washington, George, 190
Water density, sea level and, 19
Water rights, ownership of, 36
Wave height, 66
Waveland, Miss., 7, 138–39, 142, 145, 147
Wavelengths, 66
Wave period, 66
Waves and wave energy, 2, 9, 12, 13, 16, 29, 32, 55, 59, 61, 65–67, 90, 92–93, 96, 188, 249; "air hammers" created by, 33; bending (or refraction), 85–88; birthplace of, 75–76; off Cape Cod, 34; causes of, 67; and continental shelf, 70–71; forms of (how to understand), 75–78; longshore currents and, 81–83; seismic, 122–23; shallow water, 76; shape of the beach and, 84–93; storm, 74–75, 79–80, 130, 200; for surfers, 76–77; tides as, 68–71; train of, 78; tsunamis, 71, 121, 122–32; wind and, 73–75
Wave trains, 65–66
Weather trends, 19
Wellfleet, Mass., 161
West Coast, 25, 209; checklist for buying or building on, 285–92; tsunamis, 127–28. *See also* names of beaches; states
Westport, Conn., 246
West Timbalier Island, 47, 210
Whitaker and Baxter (ad firm), 244
White, Duane, 269
Whitney (landowner), 234–35
Wick Bay, Scotland, 79
Winds, 5, 12, 19, 67, 96; off Cape Cod, 34; hurricanes and storms, 128–32; velocity of, 74
Woods Hole Oceanographic Institute, 73, 257
World War II, 157
Wrightsville Beach, 108, 196, 216, 300

Yaquina Head, Ore., 309
Yarmouth, Mass., 161

Zeeland, 157–58, 162
Zuider Zee project (Netherlands), 157